D0933370

KNIGHTS OF

Darkness

Mr. Books (Pvt) Ltd.
10-D, Super Market,
Islamabad. Tel: 278843-45
Fax: 278825

To Peter Gilbert, "Loner," the Warriors of the Zendokan, the Thorns of the Black Flower, the Faithful Tigers of the Kalimarga, James Rader, John F. "Red" Johnson, James Scivens, and Carl "Groundhog" Mason.

Knights of

Darkness

Secrets of the World's Deadliest Night Fighters

Dr. Haha Lung

Paladin Press · Boulder, Colorado

Other Books by Dr. Haha Lung:
The Ancient Art of Strangulation
Assassin! The Deadly Art of the Cult of the Assassins

Knights of Darkness:
Secrets of the World's Deadliest Night Fighters
by Dr. Haha Lung

Copyright © 1998 by Dr. Haha Lung

ISBN 0-87364-971-0
Printed in the United States of America

Published by Paladin Press, a division of
Paladin Enterprises, Inc., P.O. Box 1307,
Boulder, Colorado 80306, USA.
(303) 443-7250

Direct inquiries and/or orders to the above address.

PALADIN, PALADIN PRESS, and the "horse head" design
are trademarks belonging to Paladin Enterprises and
registered in United States Patent and Trademark Office.

Illustrations by Ralf Dean Omar

Contents

Warning

Many of the techniques and devices discussed in this book are extremely dangerous and possibly illegal. Before attempting to perform any act or use any equipment and techniques discussed herein, the reader is advised to receive professional training and to ensure that he is in complete compliance with all federal, state, and local laws and regulations, and is not, in any way, endangering others. This book is *for academic study only.*

Introduction

MASTERS OF THE GAME

Good things of day begin to droop and drowse
Whiles night's black agents to their preys do rouse.

Macbeth
Act III, Scene III

From the time he clung precariously to the darkened branches of the primordial World Ash, cowering without the comfort of fire nor the knowledge of how to invoke it, man has feared the coming of night. Even after mastering flint and finding respite in his small communal campfire, still man shivered at what he imagined was salivating just beyond his little island of illumination. So man filled the world with well-lighted cities, pushing back the night inch by inch. But still they came, those scurrying and skulking corruptions that have haunted man's sleep since infancy, only now these lurkers' asylums are the skewed shadows cast by man's great towers, molded by his garish neon, animated by his guilt, unchecked by his imagination, and fed by his fear. But in every age there are those few men who find fear an unpalatable succor, warriors who instead belie the lumps in their throats, grip their weapons all the tighter, and venture forth into the night to meet Death face-to-face, some to challenge him, some to serve him.

And of those few to return, fewer still are those who will or can speak of what they have seen. Perhaps the euphemisms of those who hid in the light fit ill the experience of these returning knights of darkness. Or perhaps too few of those cowering in the light are worthy to share in the secrets the night and Death have whispered to such hardy souls. Yet the changed eyes of these reborn night warriors speak volumes to any who dare to meet their gaze and read the deeper shadows and silhouettes now etched there. Choking on the bitter tea of tradition and taboo that did so little to mask the bile of tepidity, these transformed knights laughed openly in the face of the herd. And soon the herd learned to fear this laughter, even as the night itself was feared. And so these dark knights were hunted by rabid packs of the fearful and brought to ground even as prouder beasts are laid low by scurvy jackals.

Oh, but they did not go gently into that good night. They sold their lives dearly, these knights of darkness, the putrid blood of their detractors anointing their blades, washing their sure and calloused knuckles.

Smiling and with defiant curses on their tongues, these champions fearlessly hurled themselves into that still deeper night without end, dragging scores of screaming, terror-maddened foes after them. Still, on nights when only the snores of the indolent and the ignorant mar the silence, the hearts of the young squires fathered by knights past can be heard to beat arrhythmic inside their bone cages. As the ears of their sires, their ears are tickled and tempted by antediluvian chants borne on subtle currents birthed somewhere out in that unexplored ebon, just beyond the reach of the light.

Should we marvel then, that while others draw covers tighter over their heads, sleep troubled by imaginings of foul things scratching at the doors to consciousness, that these squires slip off the blankets of malaise and from the glare of the light to rush into the arms of the night? Following in their fathers' footsteps, they tread through the darkened wood, along a path unseen and unsuspected by the fearful, to arrive unafraid at the lip of the great abyss. Fixing it with blooded eye and nostrils aflare, unflinchingly they stare into it, never fearing what might stare back.

Poised there on the edge, they vow to self and the spirit of Sire to never again fear the dark, never to cower from, or stumble blindly into the darkness, but to faithfully serve it with silence and stealth until Death beckons them, newly beknighted, into that Great Hall reserved for warriors alone. There, having taken their rightful place at this great round table, they will join Death in toast to those knights of darkness past, those in residence, and those yet to sup at the tit of Mother Night.

MASTERS
OF THE
GAME

PART ONE

CHAPTER ONE

Masters of the East

Many of the techniques used by today's Western night fighters and special forces groups come from the mysterious Far East, but east is only a direction. Standing on the east coast of China, we would see the reborn sun rise over California. Looking east from New York, we look to Europe, home of its own master night fighters. Yes, east is only a direction, but our first steps into the night must begin somewhere.

ORIGINS OF THE NINJA

When we think of accomplished night fighters, it is perhaps natural that we think of the dreaded masters of the night of medieval Japan, the ninja. Ninja have been portrayed in more movies, television shows, books, and magazines than all other cults and cadre of night stalkers combined, but what do we really know about these Japanese knights of darkness? Were they the originators of today's commonly used methods of shadow and stealth? Or did they inherit (or steal) techniques from even older night masters?

According to legend—legends the ninja themselves encouraged—Japanese ninja, more properly known as *shinobi*, are descended from half-crow, half-man demons known as *tengu*. Tengu were forest magicians, shape-shifters, and master swordsmen who dwelt in the deep forests of central Japan. Tengu ninja operated under cover of darkness, in contrast to the accepted Bushido code of the samurai, which dictated opponents fight face-to-face.

The truth of the origins of the ninja is only slightly less mythical and dramatic.

Moshuh Nanren

Centuries before the advent of the shinobi ninja in Japan, China had its own ninjalike spies, the *moshuh nanren*. The moshuh nanren plied their trade of spying and assassination for centuries, beginning in China's Warring States Period (453–221 B.C.E.). It was during this tumultuous period in Chinese history that General Sun Tzu wrote his classic *Ping Fa* ("Art of War"), which outlined all aspects of warfare, including the use of *k'ai ho* (literally "gap men," spies). According to Sun, "there is nowhere that you cannot put spies to good use."

Chinese emperors took Sun's advice to heart, establishing cadres of special spies whose purpose was to ensure for their emperor a safe and prosperous reign. Moshuh nanren spied for the emperors, assassinated rivals and traitors in the imperial court, and occasionally forestalled a revolt by making troublesome rabble-rousers disappear. To help instill fear in their enemies, moshuh nanren perpetuated the myth that they were descended from *lin kuei*, night demons dwelling in China's forests who were thought to possess mystical powers. They were masters of disguise, which helped fuel paranoia among the peasants, who never knew when the fellow peasant they were complaining to might be moshuh nanren incognito, collecting intelligence for the emperor. Accomplished knights of darkness, moshuh nanren struck in the middle of the night, dragging off suspected traitors and other enemies of the emperor or slaying them while they slept.

One of the moshuh nanren's favorite ploys was perpetuating

the belief that they possessed the secret of killing with a single touch, a technique known as dim-mak. While experts still debate whether ancient Chinese assassins actually possessed such a secret, it is a fact that many moshuh nanren, like their Japanese ninja cousins who followed, routinely used poisons and ploys known as "one-eyed snake" that were designed to mimic genuine "death touch" techniques.[1]

Whenever an emperor died or was deposed, his supporters would often be killed by the new emperor. As a result, moshuh nanren agents of fallen emperors survived by melting into general Chinese society. All these displaced moshuh nanren continued to employ the techniques and tactics of shadow and stealth that were their bread and butter. Some of these out-of-work assassins sold their skills to the highest bidders, such as ambitious warlords and rich merchants. Others founded political secret societies opposing the new emperor, and still others joined or organized criminal *t'ongs* ("brotherhoods").

Over the years, techniques of displaced moshuh nanren and, indeed, several exiled moshuh nanren masters found their way to Japan, Korea, Indonesia, and Malaysia. There they influenced the development of such night fighters as the ninja of Japan, the *hwarang* ("flower knights") of Korea, and the "nightsider" criminal gangs of Malaysia.

Ninjutsu

Ninjutsu—the art of the ninja—first made its appearance in 6th century Japan. During this period, Prince Shotoku Taishi, caught up in a bloody war of succession, hired an accomplished spy by the name of Otomo-No-Saijin. Otomo's nom de guerre was Shinobi, and it is from the written character for Shinobi that we derive the word *ninja*, literally, "one who sneaks in." That spying techniques and night-fighting tactics employed by Otomo should resemble those used by the moshuh nanren comes as no surprise, for moshuh nanren of various ilk—spies for the Chinese government, operatives from secret societies, t'ong agents—infiltrated into Japan between the first and fifth centuries, hidden in an influx of Buddhist missionaries. Some scholars have gone so far as to suggest that Otomo himself may have been a moshuh nanren agent. Others have speculated that Japanese spying techniques developed as a counteraction to Chinese spying efforts. Whatever the truth of the matter, the fact is moshuh nanren spying tactics and techniques planted the seed from which the great oak of Japanese ninjutsu grew.

A second major influence on the development of medieval Japanese ninjutsu was the destruction of several *yamabushi* (mountain warrior-monks) monasteries in the 12th century after the rise of the shoguns and the standardization of the samurai warrior class. Many of these military-trained warrior-monks were forced underground by samurai persecution and settled as farmers in central Japan, but they continued to strike back against their samurai overlords under cover of darkness. By the Middle Ages, displaced yamabushi cum shinobi had welded themselves into powerful clans of ninja. Some of these clans sold their considerable night-fighting talents to the highest samurai bidder, playing one samurai faction against another. Ninja strategists reasoned that the more time samurai spent fighting one another, the less time they would have to persecute the shinobi clans.

Over the years, shinobi ninja honed their night-fighting tactics and techniques to such a level of perfection that the mere mention of their name was enough to strike terror into even the stoutest samurai heart. So feared were the ninja that, by the 17th century, even uttering the word *ninja* became punishable by death.

Today, when we use the noun *ninja*, we picture a masked warrior in black, armed to the teeth with hidden weapons. In medieval Japan, however, *ninja* was a verb applied to the actions of anyone who used stealth; everyone from spies to burglars. During their heyday, the great shinobi ninja clans of Iga and Koga Provinces of central Japan inspired awe and fear up and down the entire country.

During these centuries of internecine samurai warfare, ninja lived high off the hog, with plenty of business opportunities. However, with the unification of Japan and the disenfranchisement of the samurai, such opportunities for ninja dropped off dramatically. As business dried up, many ninja turned their night-fighting talents to crime, helping form the basis for Japan's version of the Mafia: the *Yakuza*. Other ninja took the opposite tack and went into law enforcement. Many continued plying their dark craft under the auspices of ultranationalist secret sects, such as the Black Dragon Society. Still others joined the military, helping to train Japanese troops in night fighting up to (and since) World War II. Today, the Yakuza, as well as legitimate Japanese business concerns, employ ninja spies to gather intelligence and for other well-known ninja specialties.

Although these Japanese ninja are who we generally think of when imagining oriental night fighters, they are by no means the first, nor the only, Eastern knights of darkness.

Hwarang

The three warring kingdoms of the Korean peninsula were united under the Silla Dynasty in 668 C.E., forging a single nation. Part of the effectiveness of the Silla forces, both overt and covert, were the Silla's elite *hwarang* warriors. The hwarang were a paramilitary youth corps made up primarily of the sons of Silla elite. The hwarang, and their philosophy and martial arts training, collectively known as *hwarang-do* ("path of the hwarang"), was held up to the newly unified nation as the ideal. Hwarang-do philosophy was a mixture of Taoism, Confucianism, and Buddhism, stressing respectively, harmony with nature, loyalty, and filial piety, as well as transcendence of worldly matters leading to enlightenment.

Hwarang were devotees of Maitreya, the Buddhist messiah yet to appear. They preached that a balanced way of life was the ideal to strive for. This ideal was put into writing in 602 C.E. by the Buddhist monk Won'gwang: "Serve your Master with loyalty and your parents with filial piety. Use good faith in communications with friends. Face battle without retreating and, when taking life, be selective."

When it came to taking life, hwarang were not only selective, but were also very good at it. As warriors, they were required to master archery, horsemanship, hand-to-hand combat, and the art of stealth. During the wars of unification leading to Silla's dominance of Korea, hwarang were the vanguard of Silla's advancing army, acting as guerrilla shock troops and *sappers* (night infiltrators). They were adept at infiltrating enemy lines at night to assassinate rival commanders and sow havoc among the enemy.

In the *Silla Kukki* ("Lost Records of Silla") there are references to the effeminate, ostentatious courtly display affected by the hwarang (this in startling contrast to the ferocity of the hwarang's martial arts skills in general and their night-fighting ability in particular).

Many suspect that the night-fighting tactics and spying techniques of the hwarang originally came from the moshuh nanren, and there is no doubt that Chinese culture greatly influenced the development of Korean culture, as it did that of the Japanese. But to what extent hwarang night-fighting and spying acumen derived from the moshuh nanren may never be known.

As it did for many elite military cadres, the decades of relative peace and prosperity following the unification of Korea spelled doom for the hwarang. Today, through the efforts of martial arts instructors coming to the West, there is a renewed interest in the study of hwarang philosophy and martial arts in general and their night-fighting skills in particular.

Thuggee Faithful Tigers

Japanese ninja only seized popular imagination in the West after World War II. Prior to that, if you were writing a pulp novel or needed a shadowy assassin to portray in your movies, for example, in the original silent *Phantom of the Opera*, you picked a *thuggee* strangler.

The East Indian killer cult of thuggee stretches far back into India's history, but did not come to Western notice until the much publicized "Thug War" carried out by British colonialists in the first half of the 19th century. In the same way that *ninja* was used generically in Japan to mean anyone employing the art of stealth, likewise *thug* was widely applied to an array of killer cadres operating in central and northern India down through the centuries.

The original *thugs* were followers of Kali-Ma, a fierce warrior-goddess still worshiped in India. According to legend, Kali-Ma slew a great demon by cutting off its head. Unfortunately, from every drop of the slain demon's blood sprang another demon. To remedy this, Kali-Ma transformed nine Bengal tigers into men and taught them how to strangle these demons using a sacred cord woven from her hair. She then gave her "Faithful Tigers" and their descendants divine sanction to slay any demon, i.e., any nonbeliever. As a result, hunting parties of thuggee Faithful Tigers waylaid travelers all across the Indian subcontinent until the thuggee were suppressed by the British in the early 1800s.

What originally began as an obscure Hindu cult later grew to include bands of criminals and displaced Islamic *hashishin* assassins, and in 1829 the British created a special department for the suppression of thug gangs. Between 1831 and 1837, a total of 3,266 thugs, as well as Islamic infiltrators and criminals posing as thugs, were captured by British authorities. More than 400 thugs were hung and the rest deported. By 1860, the British declared the thuggee wiped out as a widespread problem, although bloody revivals of thuggee practice have cropped up since then.

Before the thuggee were executed or deported with typical British efficiency, interrogators picked thuggee brains clean of their night-fighting tactics and techniques. As a result, thug techniques of night-fighter stealth and sentry removal lived on in tactics and techniques taught to Britain's fledgling commando and espionage corps. Via British commandos and spies, such thug methods eventually filtered down to other Western special forces. In addition, other deported and displaced thuggee practitioners went on to teach secrets of stealth and slaughter to many diverse groups of guerrillas, terrorists, and criminals throughout the Far East.

The Nightsiders

The archipelago stretching from Malaysia to New Guinea, comprising more than 3,000 islands, is home not only to legitimate traders but also to a complex criminal underworld known as the *nightside*. Analogous to the Sicilian Mafia and the Japanese Yakuza, Malaysian nightsiders are a loose, often warring confederation of gangsters, pirates, smugglers, and professional assassins. Organized into t'ongs, these gangs control all criminal activity, from gambling and prostitution to drug smuggling. Over the centuries, specialists in spying and assassination known as *naga* ("dragons") developed within the nightside underworld.

Given the nature of the area—the constant infusion of travelers, refugees, and exiles from all over the Far East—it is not surprising to find influences in naga night fighting derived from Indian thuggee, Japanese ninja, and Chinese spies and *Triad* (Chinese mafia) assassins. Because of a large population of Muslims living on the islands, we also find influence from Islamic *daggerman* assassins among the naga, evidenced by the fact that the *kris*, a wavy-bladed dagger of varying lengths, is one of their favored killing tools. Other naga prefer a more hands-on approach and employ unarmed killing blows derived from the art of pentjak silat, a system of martial arts similar to kung fu. In addition to their armed and unarmed killing arts, naga maintain a wide variety of psychological ploys designed to strike terror into the hearts of victims before the actual dagger plunge or the blow that stopped the victim's heart. (The naga's favorite saying is "Terror is the sharpest of blades.")

Hashishins

Intelligence gathering is one of the two main objectives of a night fighter, the other being assassination. The word *assassin* comes to us from the cult of the hashishin, a fanatical Muslim sect of night-stalking medieval killers who spread terror throughout the Middle East for more than two centuries. Founded in the late 11th century by Hasan-ibn-Sabah, whom Marco Polo dubbed "The Old Man of the Mountain," the hashishin cult was an offshoot of the radical Shiite branch of Islam. All modern Islamic terrorist groups trace themselves back to this cult of killers. From their mountain stronghold at Alamut in western Persia (Iran), hashishins spread out, assassinating and controlling by intimidation anyone who dared defy the wishes of The Old Man of the Mountain. The hashishins' favorite technique was to sneak past all the security a king or potentate could muster and, in the dead of night while the official slept, place a dagger beside his head. For those few who wouldn't take this hint, hashishin *fidavis* ("daggermen") would make a return visit, stepping from the shadows of a mosque while the target was praying or again slipping passed armed guards and bolted doors to slay the victim as he slept.

Eventually, not only their fellow Muslims but invading European crusaders as well learned to fear this cult of assassins. Like the samurai of Japan, the European crusader knight who thought nothing of facing multiple foes on an open battlefield in the bright light of day cringed at the thought of a lone assassin's blade striking from the dark. So great was the crusaders' fear of these Islamic slayers that European knights took a new word with them when they returned to their northern homelands: *assassin*. The word is a corruption of hashishin, literally "one who eats hashish." Popular legend has it that these killers would get high on the drug before attacking their victims. The drug supposedly numbed them to wounds, allowing them to cut down their target before succumbing themselves. The word crusaders misunderstood as hashishin, thus assassin, was probably *Hashimite*, a Muslim term denoting a person descended from Muhammad, the founder of Islam.

To accomplish their missions and ensure that no foe of the cult would ever escape its blades, the assassin cult developed an impressive array of practices, ploys, poisons, and propaganda, most of which are still in use today. Over the years, assassin daggermen were responsible for a score of important political slayings throughout the Middle East, including the assassination of several European crusade leaders. More disconcerting still were rumors that some of the slayings of these crusaders were contract killings carried out by the assassin cult and paid for by rival crusaders. Richard the Lionhearted was even accused of having contracted assassins to kill rival Conrad of Montferrat in the late 12th century. In addition to such contract killings, several prominent members of European royalty and knight orders were implicated as members of secret societies believed to have links with, or to have been secret initiates of, the cult of assassins.

In the 13th century, the assassin cult's stronghold at Alamut was destroyed by invading Mongols, and many Persian assassins fled to India where they infiltrated the thuggee sect. Other assassins fled to European-held areas of the Levant. Some of them trained knights and bodyguards, passing on secrets of stealth and skulduggery that would eventually find (mis)use in Europe proper.

ENDNOTE

1. Ralf Dean Omar, "Ninja Death Touch: The Fact and the Fiction," *Black Belt* (September 1989).

CHAPTER TWO

Masters of the West

Down through the ages, the West has not been slack in producing its own cadres of night fighters. Some of these have a history that goes back to time unrecorded, while others have their origins as recent as the 20th century.

WOLFSHIRTS

Northern European tribes have a long history of warfare prior to the eighth century, when Viking raiders and traders began to frequent the coast of the British Isles, France, and the Mediterranean. These fierce, flaxen-haired Norse raiders were often organized into warrior brotherhoods, in many ways comparable to the *Shao-lin* warrior-monks of China and the yamabushi of Japan. The Norsemen honored bravery above all else and thought nothing of throwing themselves into battle against far superior forces. But the Norsemen also admired guile and cunning, and Norse history is replete with tales of how one warrior outwitted another. Tales abound in Norse mythology of how the gods themselves—Odin, Thor, Frey, and Loki ("the

Trickster")—used craft and deception in their dealings with mankind and with each other. Some bands of Norse warriors specialized in guerrilla tactics, mastering the art of stealth and striking in the dead of night.

Collectively known as *ulfhedin* ("wolfshirts"), several of these warrior bands more closely resembled the ninja in tactics and techniques than the general idea we have of berserk, horn-helmeted Vikings. The tradition of wolfshirts stretches far back into Northern European history, with the first written mention of them coming from the Romans. For example, warriors of the Germanic *Harii*, famed by the Roman historian Tacitus for their strength and daring in battle, cultivated the art of terrifying their enemies: "They black their shields and dye their bodies black. The terrifying shadow of such a fiendish army inspires a mental panic, for no enemy can stand so strange and devilish a sight." The Harii fought without armor, protected only by their shield and a cloak. As discussed more fully in the chapter on camouflage, such cloaks, most often a wolfskin, were used to break up the wearer's silhouette, making the wearer a more difficult target by allowing him to change his body's outline, i.e., "shape-shift."

Tattooing and body paint were also widely used by wolfshirts, both as psychological war paint and as practical camouflage. Many of these European ninja fought naked ("sky-clad") so as to be better able to tell naked friends from clothed foes when fighting in pitch darkness. Because they often fought against superior odds, wolfshirts mastered the arts of stealth, removing sentries and infiltrating enemy camps and villages before the enemy realized Death had come calling.

The envy of any modern SEAL operation, wolfshirts would drift silently down rivers, gliding ashore just before dawn to strike enemy villages or gilded English monasteries. With feudal power becoming more centralized, many wolfshirt bands disbanded, but some adapted and became spies for ambitious princes or trainers for royal bodyguards. Still others sought more fertile ground overseas, hiring out as enforcers and mercenaries for kings and merchants as far away as Constantinople.

KNIGHTS TEMPLARS SHIRKERS

During the Crusades, various groups of crusading knights orders came into being. The most well known were the Knights Templars, founded in 1118. In a relatively short time, the Knights Templars went from being an impoverished order to being the richest in Europe, boasting castles and estates stretching from Paris to the Levant. As they grew more powerful, the Templars inevitably acquired powerful enemies who needed watching. With incredible foresight, especially in an age of internecine warfare and petty nationalistic loyalties, the grandmasters of the Knights Templars set up Europe's first international intelligence network with fielding operatives in every European kingdom and principality. These Templars *intrigant* (a French term meaning "one who engages in intrigue") agents were also known as *shirkers*. In medieval times, the word *shirk*, today synonymous with evading the performance of a duty or obligation, meant to go stealthily or to sneak and aptly describes the intent and actions of Templars shirkers.

Shirkers were expected to master the twin techniques of *shirk and dirk*—stealth and assassination. Their motto: "an ear to every crack, a dirk through every crevice." They spied on and removed through bribe, scandal, or assassination anyone who posed a danger to the order. To accomplish their goals, shirkers defied the European knights' rule of chivalry by employing secrecy and stealth to accomplish their goals.

Some claim that the shirkers learned their secrets of shadow and stealth from the dreaded cult of assassins.[1]

After the Templars' order was suppressed at the beginning of the 14th century, surviving Templars scattered, some helping to found Masonic secret societies. Some Templars shirkers helped disseminate techniques of shadow and stealth by taking work as mercenaries, spies, and martial arts instructors.

European commanders slowly learned the value of maintaining their own elite cadres of night fighters. In the mid-1500s, Julian Romero, one of the most renowned captains in the armies of King Philip II of Spain, created the elite *encamisada* ("in your shirt") squadron. In an age of heavy body armor, encamisada traveled light, trading heavy armor for light clothing and the barest of armor (a steel corset and helmet) in order to allow squads to march faster and farther than any other unit. When they did wear breastplates, they always wore a shirt on the outside so they could tell friend from foe at night. Each was armed with a sword, poniard, pistol, and arquebus. No encamisada rode a horse, as horses make noise.

So successful were these knights of darkness that encamisada was soon being used as a synonym for a swift, stealthy, nocturnal raid with a small force. The encamisadas' most famous, or infamous, mission was their attempt to kidnap or kill Holland's Prince William Orange, who had revolted against the Spanish overlord of Belgium and Holland. Learning that William was holed up in a tent in the center of an encampment at the village of Herminhny, Romero picked a moonless night to land his small force. Silently crossing previously reconnoitered terrain, the encamisadas easily removed sentries and penetrated the encampment before the alarm could be sounded. Ironically, Prince William escaped capture only because the barking of his pet spaniel had roused him minutes before. Although they missed their main objective, Romero's men killed more than 600 of William's guards while losing only 60 of their own.

Napoleon also had a special corps of barricade sappers known as "The Forlorn Hope," whom he would call upon to break the stalemate of a siege by infiltrating enemy lines. The Forlorn Hope often threw themselves into suicidal charges in broad daylight, but were just as feared for their ability to penetrate enemy lines and perimeters in pitch darkness, cutting throats and preparing the way for more traditional troops. (Note: Sun's *Ping Fa* was introduced to the West by a French missionary shortly before the French Revolution. We are left to speculate how much of an influence Sun's strategy, especially the chapter on the use of spies, might have had on France's future emperor.)

EARLY AMERICAN NIGHT FIGHTERS

Contrary to John Wayne movies, most Native American tribes and warrior societies had no qualms about fighting at night. In fact, like most native peoples, American Indians saw nighttime as the best time for stealing horses and slaves—and for cutting throats.

The night-fighting ability of the Indian terrified European settlers. The Indians' skills were light-years ahead of the colonists' because, prior to the coming of the white man, Native Americans had amused themselves by raiding rival tribes who were equally skilled in the arts of stealth and night fighting.

As a result of centuries of such internecine warfare, the American Indian learned not only to use techniques of stealth himself, but also how to guard his own throat and horseflesh against rival knights of darkness. A brave learned early on how to move silently through the bush; his war paint was both decorative and functional as camouflage. Like the wolfshirts of Europe and the leopard cult killers of East Africa, Native Americans often wore animal cloaks—buffalo, wolf—to disguise their form and help them better blend in with the smell and feel of their surroundings. Fringed buckskin and feathers also helped break up a brave's silhouette. His weapons—the bow and arrow, tomahawk, lance, and knife—all lent themselves to stealth.

It is a pity so few Europeans took the time to learn the techniques of stealth and night fighting used against them. One wonders how many lives might have been saved had our armed forces begun training special forces in night fighting a hundred years before they finally did.

From the American Revolutionary War comes an overlooked story of night-fighter tactics.

British troops were already shaken after repeated encounters with rebel colonists who steadfastly refused to play by the rules of civilized European warfare. Used to fighting comparable European troops on open battlefields, British commanders were at a loss in the face of the colonists' Indian tactics.

Camping after a hard day's work chasing rebel shadows, several British soldiers sat around their fire complaining loudly of how cowardly the rebels were. Silently, two cloaked figures detached themselves from the shadows. Their night vision destroyed from staring into the campfire, the surprised British soldiers blinked in an effort to identify these two cloaked figures.

"I understand you believe all colonists to be cowards," the first cloaked figure said, his tone making the seated soldiers glance nervously in the direction of their muskets, which were stacked far out of arm's reach.

The first redcoat was dead—a pistol ball between his eyes—before any of the other soldiers could react. The second shadow followed suit and discharged two pistols that had miraculously appeared in his fists, dropping two more redcoats. Soldiers in the immediate vicinity fled in terror, even as other British soldiers, with firearms, joined the melee.

Having discharged a brace of pistols each, the cloaked figures fought on with bayonets, then with tooth and nail, bringing down a score of British soldiers before being killed themselves.

It is a shame that the names of these two intrepid knights of darkness were never recorded, save on the heart of Mother Night.

Slowly, America's fighting men learned the importance of night fighting. The American Civil War saw many groups of Confederate guerrillas, all adept night fighters. The most accomplished of these were Mosby and his Partisan Rangers. Mosby intuitively understood the advantage fighting at night gave a numerically smaller force. As a result, Mosby perfected the science of guerrilla warfare, with special emphasis on the art of night fighting. Once described as "a caped specter on a ghostly grey charger," Mosby won the title "Grey Ghost" in recognition of his ability to disrupt Union communications, capture enemy officers, and disappear back into the night without a trace.

ENDNOTE

1. Dr. Haha Lung, *Assassin! The Deadly Art of the Cult of the Assassins*, (Boulder: Paladin Press, 1997).

CHAPTER THREE

Modern Masters of the Night

WORLD WAR II COMMANDOS

From the sneak attack on Pearl Harbor to the secrecy surrounding the Manhattan Project, mastery of deception and stealth was the key to success in World War II. Anyone who doubts this need only look at the degree of deception the Allies resorted to in order to disguise the preparation and execution of plans for the Normandy invasion, Operation Overlord. We also need only look at the large numbers of commandos dropped into Normandy prior to the actual invasion to see how the use of guerrilla tactics and the deployment of accomplished night fighters had finally earned a place of respect in the minds of Western military commanders.

Dropping from the night sky over Europe by parachute or silent gliders, Allied commandos moved through the dark, blowing bridgeheads and preventing the Axis from responding to the Allied landing. It is surprising that it took this long for Western military strategists to learn respect for the knights of darkness, since America already had several examples of how effective guerrilla warfare and special forces night fighters could be. These examples came from Native Americans, Revolutionary War fighters such as Ethan Allen and his Green Mountain Boys, and "Swamp Fox" Francis Marion to Confederates Col. John Singleton Mosby and William Quantrill.

Three factors helped make special forces possible during World War II. First, unlike wars of previous centuries, or even World War I, World War II had more innocent bystanders killed. Many Europeans, their outdated armies easily overrun by superior Axis firepower, suddenly found themselves living in a battle zone. Many of these trapped civilians actively resisted the Axis by forming such guerrilla cells as the Marquis, realizing that one determined person applying a crowbar to a railroad track or a saw to a strategically placed tree could halt an entire column of Axis troops and equipment.

Once the effectiveness of such impromptu guerrilla units became obvious, Western military leaders began authorizing the creation of special forces designed to operate behind enemy lines.

A second major factor contributing to the rise of World War II special forces was the development of smaller weapons and more powerful explosives, allowing smaller, fast-moving units to deliver the same impact as larger, slower-moving groups.

The third factor was the development of parachutes, gliders, and scuba gear, all of which allowed these better-armed, specially trained commandos to silently infiltrate enemy lines, where they could do the most damage.

As noted, the British developed special commando units and trained them in night-fighting and sentry-removal techniques that were gleaned from their suppression of Indian thugs and other guerrilla fighters within the British Empire. The British also put effort into their espionage and counterespionage departments, dropping agents into Axis-held territory to perform acts of sabotage and assassination and to help train partisans.

Likewise, the United States organized such commando units as Darby's Rangers and Merrill's Marauders. In the Pacific, Merrill's Marauders went up against Japanese troops trained in night fighting by genuine ninja masters. Surviving members of these initial Western commando units learned valuable lessons that would be incorporated into the formation of future Western special forces.

German commandos were also active throughout World War II, from sapper units that were a vital part of the blitzkrieg to units of *werewolf* counterguerrillas organized near the end of the war. Trained in guerrilla tactics and night-fighting techniques, these werewolves were assigned the duty of slowing the Allied advance and carrying out a protracted guerrilla campaign designed to reestablish Nazi power. (Though limited in scope and successes, many of these counterguerrillas continued to fight on for the "thousand-year Reich" long after their leader lay dead in a bunker in Berlin.)

MODERN SPECIAL FORCES

After World War II, Western military leaders toyed with the idea of establishing permanent special forces units that would be beyond existing commando groups, but it took the Korean War to make Western commanders see their importance, especially those adept at night fighting. During the Korean War, United Nations troops came up against Chinese irregulars descended from the moshuh nanren skilled in night-fighting techniques. Fortunately, Republic of Korea (ROK) troops, schooled in night-fighting tactics and techniques of the hwarang, were able to hold their own against these Chinese guerrillas. Later, these latter-day hwarang warriors would pass along some of their combat skills and night-fighting acumen to their occidental allies.

Out of the frying pan and into the fire—straight from Korea to Vietnam—the United States military finally saw fit to authorize the creation of special forces units.

The Green Berets

In 1952, the U.S. Special Forces ("Green Berets") were born. Whereas the existing U.S. Army Rangers, established during World War II, tended to work in larger units closer to friendly territory and for shorter periods, teams of Green Berets were specially designed to be self-sufficient, penetrate deeper into hostile territory, and stay there for longer periods. The Green Berets were mandated to live with, learn from, and train indigenous guerrillas in enemy-occupied areas, and were championed by President John F. Kennedy.

Green Berets specialize in unconventional methods of operation, not the least of which are night-fighting tactics and techniques. Like medieval Japanese ninja, Green Beret teams work in three groups, and each group has a special role to play in an operation.

Group one's specialty is surveillance and intelligence. They are responsible for initial reconnaissance and the gathering and interpretation of intelligence. The second group is the security element, which is responsible for getting the team to the mission site safely (ingress) and for covering the team's escape (egress) from the area. Group three is the assault element, the members of which are responsible for carrying out the actual sabotage, kidnapping, or assassination.

Special forces in general, and the Green Berets in particular, emphasize the development of the individual and his ability to think on his feet and improvise in a crisis, as opposed to blindly following fixed orders. In Vietnam, Green Berets studied the tactics and techniques of the Vietcong and then turned those tactics against them. These Special Forces knights of darkness lived in the bush, learning which sounds and smells were natural and which were not. They mastered the art of silent movement through the black jungle, waiting, unmoving—sometimes for days—to spring an ambush before quickly disappearing back into the night. For the first time, "Victor Charlie" learned to fear the coming of night.

The Vietcong

In 939 C.E., the Vietnamese people drove out their Chinese overlords. In 1288, Vietnamese warriors defeated the forces of Kublai Khan, Mongol emperor of the Chinese Yuan Dynasty and grandson of Genghis Khan. It took the French imperialists 26 years (1858–1884) before they could confidently declare Vietnam subdued. They were wrong.

In 1945, a frustrated Japanese occupation force pulled out of

Vietnam, and in 1954 after the fall of Dien Bien Phu the French were finally convinced to do the same. By 1965, when history-dyslexic Americans first began wading ashore in Indochina in large numbers, the Vietnamese already had more than 500 years of constant fighting—most of it at night—to sharpen their *punji* sticks. Nearly 20 years later, when it was all said and done so far as the Americans were concerned, booby traps would be responsible for 17 percent of American wounds and 11 percent of American deaths during the Vietnam War. Operating in small bands as befit a guerrilla force, Vietcong sappers preferred hit-and-run night fighting to more conventional daytime battles. Masters of moving silently through the night, intimate with the sights, sounds, smells, and terrain of their homeland, the Vietcong defied conventional forces. Only Special Forces had any chance of interdicting them.

During the siege of Dien Bien Phu, Vietnamese sappers patiently dug their way under French perimeters to strike deep into the heart of the compound. And long before the Americans arrived to find black pajama'd shadows waiting down every trail, Vietnamese sappers had mastered the art of the tunnel, both as shelters for survival and as the best means to literally undermine an enemy. During the American occupation, tunnels were dug near, into, and completely around American bases. In places, these access tunnels allowed sappers to come and go from American installations at will.

Some of these Vietnamese tunnel dwellers literally lived underground, wrapped in darkness for weeks at a time. When they did emerge, it was only at night, the topside world where they plied their dark craft a hundred times more lighted, to their eyes at least, than the pitch black they had left below. Accomplished knights of darkness all, men and women alike, these sappers slithered from their hidden tunnels, striking light-intoxicated enemies, before again disappearing, literally swallowed up by the earth. Even when one of these vast tunnel systems was uncovered, few Americans cared to crawl down into them. Fewer still were those who crawled back out unscathed.

American Tunnel Rats

After several years and many casualties, the Americans got smarter and created special squads for dealing with enemy tunnels. These American "tunnel rats" are the most unsung heroes of Vietnam.

Specially recruited for size and nerve, tunnel rats boldly crawled down into dark enemy tunnels with nothing more than a pistol, a knife, and a flashlight, the last of which they seldom used. Deep in the tunnels, some stretching for miles and going down three stories, tunnel rats engaged in gunfights, knife fights, and hand-to-hand combat in absolute darkness with an unseen enemy. Around every bend in the tunnels lay booby traps: explosives, shit-covered punji stakes, strategically placed venomous snakes and insects, and many other surprises. Yet these diminutive American knights of darkness knew no equal when it came to close-quarters combat in cramped, dark tunnels. No equal, that is, save those who had dug the tunnels in the first place.

Spetsnaz

Soon after coming to power in 1917, the politico-military leaders of the newly founded Union of Soviet Socialist Republics established an intelligence network whose job it was to gather information on enemies of the Bolshevik state, both inside and outside the Soviet Union. Part of this fledgling Soviet intelligence apparatus was the Diversionary Intelligence Service, a shadowy group responsible for carrying on secret activities, against both Soviet citizens and external anti-Soviet forces. As early as 1918, field agents, aerial photographs, and other types of intelligence gathering were being used in the Soviet Union.

One unit of the Diversionary Intelligence Service was formed from the best cavalrymen in the army, and it carried out deep raids behind enemy lines dressed in the uniforms of the enemy. These deep raiders infiltrated enemy positions, taking and torturing prisoners—primarily enemy officers—for information. So successful was this special unit that similar units were formed. Eventually all such units became known as Spetsnaz, a generic term meaning "specially assigned forces."

Today, all special forces units within the Russian military are still called Spetsnaz.

Under the Russian military system, each military district and group (land combat group or naval combat group) consists of "armies." Each army has its own Intelligence Department (RO), and each RO, in turn, fields its own Spetsnaz troops. These forces fall into two categories: commandos and agents.

Spetsnaz commandos are comparable to the U.S. Army Rangers and are trained in all aspects of combat operations, particularly night ops. In times of war, Spetsnaz commandos parachute behind enemy lines to destroy specific targets. Organized into brigades of 1,300 men each, Spetsnaz commandos are trained night fighters, equipped with the latest in high-tech equipment, including night vision scopes and listening devices.

Within each Spetsnaz brigade are 115 or more highly trained intelligence agents. Fellow soldiers, and often their commanders in the same brigade, do not know who these special agents are, since the agents are not openly identified as such. Instead, they are given "cover assignments" within the brigade as martial arts instructors, marksmen, or Olympic hopefuls-in-training and, therefore, are to be left alone.

The anonymity of these specially trained, strategically placed agents serves two purposes. First, it allows them to keep tabs on their fellow soldiers and on army commanders so that loyalty is guaranteed. (In the event of a coup attempt, Spetsnaz agents are already in position to capture or kill rebel commanders. The attempted coup against Mikhail Gorbachev in August 1991 was foiled by the mobilization of Spetsnaz forces loyal to Gorbachev.) Second, keeping their identities secret prevents these agents from being compromised, i.e., killed or enticed to turn traitor by enemy spies or coup commanders.

When necessary, in true James Bond fashion, these agents parachute into hostile territory to carry out specific clandestine operations; they are the ninja of the Russian forces.

Adept at stealth, assassination, kidnapping, sabotage, and generally creating havoc in the night, these agents operate in teams of five. When larger operations are called for, three teams are brought together to form a 15-man group. In times of open hostilities, these Spetsnaz agents have a mandate that allows them to operate deep behind enemy lines, kidnapping and torturing information from and assassinating civilian and military leaders. Unlike their commando counterparts who wear uniforms (black coveralls for night missions and other terrain-appropriate camouflage), Spetsnaz agents have no specific dress. Instead, they adapt their clothing to blend in with their surroundings (in most cases, urban dress). Under threat of a nuclear face-off, they are trained to infiltrate and destroy nuclear facilities and kill nuclear support personnel. A Spetsnaz anti-nuclear team will gladly give their own lives to disable or destroy an enemy missile capable of killing tens of thousands of their fellow Russians. In addition, each Spetsnaz operative accepts that he will be killed by his own comrades, or be expected to kill himself, should he become wounded and unable to travel.

Since Spetsnaz agents often operate in urban areas, they familiarize themselves with the language and customs of the area of operation.

Rather than outfitting themselves with all manner of exotic weaponry, Spetsnaz agents are careful to carry the same types of weapons their enemy carries. This not only allows them to blend in more easily, but ensures that they can depend on the (dead) enemy to supply them with ammunition.

Specialists in stealthy night perimeter infiltration and sentry removal, all Spetsnaz are experts in the Russian hand-to-hand combat art of sombo. More than one report filtering out of the tight-lipped Soviet Union had Spetsnaz agents practicing sombo death blows on condemned criminals and other prisoners.

A typical Spetsnaz special agent operation took place in 1975, shortly after the outbreak of civil war in Lebanon. Word was sent back to Moscow that a ranking Soviet attaché had been seized by a radical Islamic militia faction. After gathering intelligence on the faction, a team of Spetsnaz agents was dispatched to Beirut. Within three days, the body of the brother of the leader of the kidnappers was dumped on the street in front of the militia's HQ. The man's throat had been cut from ear to ear and his testicles shoved into his new smile! The Soviet attaché, shaken but safe, was released the following day. Since then, not a single Soviet (later Russian) citizen has been accosted in Lebanon—or anywhere else in the Middle East!

Similar stories have come out of previously Soviet-occupied Afghanistan and out of former Eastern Bloc countries of swift and terrible responses and of Spetsnaz-sponsored disappearances in the dead of night.

With the recent easing of tensions between East and West and the subsequent collapse of the Soviet Union, it remains to be seen what will happen to the former Soviet Union's elite Spetsnaz. No one knows whether these special forces will be kept intact as part of the new Russian army or whether Spetsnaz will be disbanded and these Russian knights of darkness left to sell their considerable deadly talents to the highest bidder.

Desert Stormers

During Operation Desert Storm in 1991, several involved nations supplied special forces to the fray. Though numbering only 9,400 out of the Allies' 500,000 men and women in the Persian Gulf, special forces played critical roles in the operation. For example, Navy SEALs slipped ashore in the night to set off diversionary explosives designed to pull Iraqi troops away from genuine targets. Groups of Green Berets were inserted deep behind Iraqi lines. Supplied with dune buggies and motorcycles specially designed with silenced engines, the Green Berets were tasked with gathering intelligence on Iraqi troop deployment and pinpointing SCUD missile sites. The most successful of these groups was a combined "fusion cell" made up of British Special Air Service (SAS) saboteurs and members of Delta Force, a U.S. Special Forces unit so elite that the Pentagon still refuses to acknowledge its existence. Over a two-week period, this fusion cell succeeded in destroying more than a dozen SCUD sites.

CHAPTER FOUR

Criminal Night Stalkers

Criminals literally live in the shadows. Think local street punks don't know what side of their favorite alleys is in shadow in the morning but not in the afternoon? Think rapists don't know every deep shadow in the local parking garage? Who shot out those streetlights, kids playing around with a BB gun or professional burglars needing more dark in which to ply their craft?

—Dirk Skinner
Street Ninja

Whereas *night stalker* is a pretty easy term to define, the word *criminal* changes with time and place. In medieval Japan, any criminal, spy, or other night skulker using stealth in his activities was labeled ninja. Hideyoshi Toyotomi, the commoner who succeeded in unifying Japan in 1590, started his rise to ultimate power as a member of a band of ninja burglars. Likewise, the terms *thug* and *mugger* both came down to us from the thuggees. Today, thug and mugger apply to a variety of criminals, although at one time they referred specifically to killers of this obscure, albeit deadly, religious cult.

According to *Encyclopaedia Britannica*, by the start of the 19th century it is estimated that one out of every 22 people living in England was a professional criminal, and as we have already seen, it is not unusual for military- or religious-oriented knights of darkness to degenerate into out-and-out criminal bands, plying their trade purely for gain rather than out of political, patriotic, or religious motivation.

It is the nature of the criminal to work in secret, preferably under the cover of darkness; thieves, burglars, and psychopathic night stalkers all use the cover of night to hide their skulduggery. The more quietly criminals move, the less they are noticed, the more they can meld with the shadows, and the longer their unsavory career. But whereas honest citizens need only guard themselves against criminals, professional criminals must not only adopt and adapt methods of stealth to cloak themselves from the police, but must also learn the ways of stealth to guard their ill-gotten gains from their fellow criminals. (No honor among thieves, remember?) For instance, drug smugglers, adept at infiltrating borders patrolled by law-enforcement aircraft, radar, sonar, and ships at sea, must also guard themselves and their contraband against rival drug runners and rip-off artists.

A while ago, a tunnel that cost smugglers millions to construct was discovered running under the border from a home in Mexico to a warehouse on the U.S. side, a classic example of perimeter penetration. Such smugglers are masters at hiding drugs and other contraband by using false bottoms and skewing visual perception to pass contraband through inspection points and police perimeters. Other smugglers have had the vocal cords of their guard dogs severed, to create "stealth pit bulls" capable of ripping into a rival dealer or robber without making a sound.

Economically motivated burglars must quickly master stealth and surreptitious night entry (or go to jail). With an increase in the sophistication of home alarm systems, successful break-in men are also becoming more sophisticated. Other home invaders, psychopathic rapists, and killers with human targets in mind rather than financial gain also utilise stealth and cloak themselves in the night. Such night stalkers often soften up a targeted neighborhood or house by first disabling lighting and cutting communications and power lines.

As we will see, the reconnaissance techniques that these night stalkers use in sizing up a neighborhood or home for invasion are the same methods special forces and others use in sizing up the perimeter of an enemy camp for infiltration.

MASTERING THE GAME

PART TWO

CHAPTER FIVE

Mind-Set

Tricks well mastered are called techniques. Techniques half learned are merely tricks.

—Ralf Dean Omar
Death on Your Doorstep

Having studied the times and terrors of past and present knights of darkness, we now move on to learn and master the tactics and techniques that will make us the equal of those past and present masters of the night. Whether our motivation for learning these methods of stealth and night fighting is for self-protection or for self-advancement, it is doubtful we will be the same at the completion of our study. Mother Night never leaves you where she finds you.

The only requirement for such study is that we give ourselves fully to concentration and effort. Night is a harsh mother, enfolding us in loving, protective arms one minute, crushing the life from us and casting us aside the next. If we cannot win her love through devotion to her shadows, let us at least earn her respect through our attention to her dangers.

In any serious and potentially dangerous undertaking, before taking that first step we must first set our minds to the proper attitude of attention and dedication that are required to master the undertaking.

Fear of what lurks in the night is the most basic of fears. Other fears that paralyze us and keep us from actualizing our full abilities are of pain, failure, and death. All too often we focus on what we fear and feel overwhelmed. We ask ourselves how we can possibly hope to fight so many fears at once. The answer is that, in truth, we have but one foe—fear—and fear can be turned from foe to friend.

For our journey into night, fear is our first and final challenge. Only after we have put fear in its proper perspective can we begin to master those mental skills that will open the night to us.

FIGHT THE FEAR

Fear can kill by either preventing us from acting or causing us to panic and act rashly. When a psycho attacks us with a knife, we feel genuine fear. To fear in such circumstances is understandable and even desirable, provided that fear does not paralyze us with inaction.

The first thing to understand is that fear isn't always bad. We voluntarily ride roller coasters, bungee jump, and sky dive, even though these kinds of activities place us under stress and induce fear in us. Likewise, horror movie fans enjoy having their pants scared off by every new *Texas Chain Saw Massacre*-type movie.

This kind of stress and fear is considered fun and exciting, since it takes place under controlled circumstances. To get fear under control, we must turn it into excitement. So long as we can reasonably control the level of fear we feel, fear is exciting and fun and we enjoy testing ourselves against it. Unfortunately, when we actually find ourselves confronted by a chain-saw-wielding psychopath, it is a little harder to keep our fear at a controlled level where it is still considered fun and exciting.

Experienced street predators see dangerous situations, such as a mugging, as fun and exciting because the situation is of their own making. Such street punks have a plan, a script, and know ahead of time what is going to happen. This gives him a feeling of control.

Fear + control = excitement.

The secret to mastering fear, thus turning it from foe to ally, lies in our realizing how all too often we misinterpret our body's signals that it is ready to fight for signs of fear.

The human reaction of fear has its origins in our primitive flight-or-fight reaction, the instinct that readies our bodies for action. Knights of darkness train to recognize these signs as those of readiness, rather than mistakenly identifying them as signs of weakness and cowardice, i.e., fear. When faced with danger, we can either fight off the threat or flee from it. Readying the body for flight or fight, the brain signals the body to shut down any functions not immediately needed for survival. Faced with a perceived threat, we get butterflies in the stomach, become flushed, and break out in a sweat, and often our limbs begin to shake. The untrained mistake these reactions for signs of fear. Nothing could be further from the truth!

Butterflies in the stomach are produced when blood rushes away from nonessential organs, such as the stomach and intestines, to the arms and legs that are needed for either running away or fighting. As warm blood rushes to these extremities it heats muscle and flushes our skin, and we break out in a cold sweat, which makes the body more difficult to grab. As additional blood-energy floods the muscles of the arms and legs, this excess energy, if not expended, produces trembling, shaking, and knocking knees. This is not a sign of weakness or cowardice, but rather a sign of readiness.

Since saliva is used for digestion and not for flight or fight, the mouth stops producing saliva, resulting in dry "cotton mouth." A person may turn pale as a result of blood flowing away from extraneous facial muscles. An increase in the brain's neurotransmitter activity enhances alertness. Pupils widen to let in more light and increase peripheral vision. Hair on the back of our necks rises, and we get goose bumps (both carryovers from less evolved times when the raising of our thick fur made us look bigger and more threatening).

Failure to recognize these natural reactions as signs of readiness, as opposed to signs of cowardice and weakness, can paralyze us with fear.

According to Dirk Skinner in *Street Ninja: Ancient Secrets for Mastering Today's Mean Streets*, medieval ninja knew a secret about fear that helped them turn fear from a foe into a friend: fear never arrives! Whatever we fear is always in the future. We fear what will happen to us when and if we venture out into those mean streets and wonder how much it will hurt if an attacker hits us.

What we fail to realize is that both of these events have yet to happen; they are in the future. Rather than live in fear of a threat that might never materialize, we should prepare ourselves mentally and physically to be the equal of any threat scenario that does appear. Once we become confident of our ability to deal with any threat, our fear diminishes. When—*if*—the call to action does come, we will not be found wanting: "Once the threat is upon us, once the wolves are at the door and the beast is at our throat, there is no time to fear, only time to react," as Skinner says. As for fearing the night itself, we must not view the darkness as an enemy, but as a friend—a friend we all need to get to know better: "In the final analysis, the same night that hides your enemy can shield you. The scale will be tipped in your favor only if the sword of your determination is sharper than your enemy's fangs," is how Skinner puts it.

Recapping, not all stress and fear is bad: "Studies on military recruits, for example, demonstrate that greater combat exposure [i.e., stress and fear] leads to enhanced self-esteem and coping skills. Regular challenges provoke enhanced stability, sociability, responsibility and courage. Moderate amounts of uncertainty and unpredictability are associated with increased senses of pleasure, hope and well-being," as David Kipper wrote in "Stress for Success" (*Prime*). In other words, a little bit of fear is healthy.

Healthy fear urges us to caution and can save our lives. Left unchecked, however, fear can become chronic, paralyzing us in moments of danger, incapacitating us for life.

To combat stress and fear, knights of darkness master the disciplines of relaxation, concentration, meditation, and visualization.

TECHNIQUES OF RELAXATION

An open, relaxed hand can caress, grasp, and, when needed, close into a striking fist. Tightly clenched, a hand can only be used to strike. A tense mind is like a clenched fist: mental fears and tension prevent us from being fully functional. Knights of darkness train to relax in order to free their mind to more quickly and correctly process information pertaining to a threat. By cultivating the ability to relax their mind, these shadow warriors are able to instantly choose from a myriad of options and possibilities, a skill vital to survival and success. A relaxed mind, in turn, releases the body's full ability.

Learning to Breathe Correctly

Relaxation begins with learning to breathe properly. Many people go around acting like life literally stinks, being satisfied with shallow gasps rather than breathing deeply of life and taking in a full measure. Most of us, in fact, breathe shallowly and irregularly.

Learning to breathe correctly is no mystery. Simply inhale as if smelling a pleasant aroma—a flower, perfume, or a favorite food—and try to get as much of the aroma as deep into your lungs as possible. Draw in a full draught of air and hold this deep breath for a mental count of three before releasing it with a heavy sigh. Allow your breath to flow out of you at a natural rate, taking your tension with it. After completely emptying your lungs, draw in another full, deep breath. Continue this breathing exercise until you feel all tension drain from your body. (Meditators often use incense to encourage this type of deep breathing/tension release.) Whenever possible, take a few minutes to monitor your breathing by practicing this technique.

Free-Word Relaxation

Each time you relax by sighing out the tension you have been holding for the three count, mentally repeat the word *free*. Repeating free each time you relax conditions your mind to trigger the body's relaxation response any time or any place you mentally repeat it, even in times of danger.

Martial artists often use a variation of this free-word technique. During martial arts practice, each time they throw a kick or punch or break a board, they either verbally or mentally shout a "trigger" word, such as "*Kiai!*," that their mind and body soon come to associate with the release of energy. Even in times of actual danger, shouting this trigger word releases the martial artist's energy.

The Two-Step Relaxation Exercise

This two-step relaxation exercise can be done sitting or lying down. The first part of the exercise is designed to relax you. The second part then infuses you with energy.

Step One: Lie down on a comfortable but firm surface, feet apart, hands at your sides. Take three deep breaths using your free-word technique. Now, curl your toes inward, tensing them for a count of three before releasing the tension and allowing the toes to relax. Repeat by focusing in turn on your legs, groin, abdomen, hands, arms, chest, neck, and finally your facial muscles, until your whole body is relaxed. Coordinate the tensing of your individual body parts with your breathing in and releasing of tension with your free-word exhalation.

Step Two: Having released the tension from your body, rub your palms together until you feel the heat of friction. With fingers slightly splayed, press your left palm just above your navel. While free-word breathing, make yourself aware of the rising and falling of your abdomen. As you breathe in, concentrate on drawing the breath down deep, to the level of your hand. Feel the warmth radiating out from your abdomen, filling you with energy.

Note: When unable to perform this full body relaxation,

place your left thumb on the inside of your right wrist where the wrist meets the palm, the remaining fingers of your left hand pressing against the outside of the wrist. Massage your left thumb in a counterclockwise motion for 30 seconds before switching to a clockwise motion for another 30 seconds. Repeat on the left wrist with the right hand. Concentrate on your free-word breathing.

TECHNIQUES OF CONCENTRATION

Awareness and Focus

Learning to concentrate and focus our attention allows us to extend our senses beyond their normal range of effectiveness. For instance, our brains routinely filter out background noise and nonthreatening distractions. However, in a room packed full of cackling people, our ears instantly zero in on that person, far across the room, who just mentioned our name. While we may not be consciously aware of every conversation in the room, at a subconscious level, our brains hear every word spoken in the room. Likewise, our brains see and record everything that passes before our eyes; every odor is detected and filed away. However, since being deluged with all this sensory input at one time would drive us crazy, our brains only bring to our attention those words or out-of-place sights and sounds that either interest us or are seen as a possible danger. Unfortunately, this filtering ensures that the average person misses out on much of what is going on around them.

Japan's greatest swordsman, Miyamoto Musashi, author of the medieval classic *The Book of Five Rings*, browbeat his students to "pay attention even to trifles." Such attention to small details requires focus and concentration.

Nature has equipped us with all the senses and instincts we need to survive. Unfortunately, nurture (society) has done its best to dull these senses and breed these instincts out of us. Ever had a feeling that you were being watched? Ever caught a friend playfully trying to sneak up on you? Did you dismiss the former as paranoia and the latter as luck?

Recently, researchers at the California Institute of Technology discovered that the brains of birds and fish contain crystals of the mineral magnetite (Fe_3O_4). This mineral acts as a biological compass to guide birds during migration and allow fish to find their way upstream to spawn. This same mineral has been found in humans and is believed to account for the electromagnetic (EM) field that extends out from and surrounds the body. Whenever another person's EM field comes into contact with our own, we "feel" it. Whether we acknowledge or ignore this feeling is a matter of awareness and training.

This feeling is known as *peripheral body sense*. Knights of darkness fine-tune their peripheral body sense awareness until they can sense when someone is watching them, even from a distance. In most people, awareness of their peripheral body sense is blunted, but knights of darkness sharpen their awareness of their peripheral body sense through the disciplines of relaxation, concentration, meditation, and visualization.

Concentration Exercises

Using your free-word to relax, close your eyes and listen for the sound that is farthest away. At first, it may be someone moving around in an adjacent room, birds chirping outside on the window ledge, or cars passing on the street. Now, *will* your hearing outward, beyond the farthest sound you hear. Practice "hearing" with your entire body. Before long, you will notice that thunderstorm rumbling in, that jet high in the clouds, and other faraway sounds that would normally be missed. (Martial artists throw practice kicks to head level because, if they can forcefully kick to head level, they will be able to kick with power to any target below head level. In the same way, extending your powers of hearing and concentration out beyond the most distant sound will help you become more aware of your immediate surroundings.) Accomplished knights of darkness know that to survive in the night you must listen with more than just your ears and see with more than your eyes.

TECHNIQUES OF MEDITATION

Knights of darkness use various techniques of meditation, some that calm the mind, some that help overcome fear, and others that stimulate strength. What all meditation techniques have in common is that they first relax the body and then concentrate the mind. Advanced meditation techniques employ visualization as well.

The Benefits of Meditation

For knights of darkness, meditation is not escapist mysticism. At the simplest physical level, meditation techniques using mantra chanting vibrate the mandible (jawbone), which in turn vibrates the cranium (skull), which vibrates the brain fluid in which the brain floats. Depending on the particular frequency of the vibration (frequencies vary with different mantras), this type of meditation practice can have either a calming or a stimulating effect on the brain.

According to recent research done at the University of Massachusetts Medical Center, regular meditation increases levels of melatonin, a hormone secreted by the pineal gland that helps to regulate sleep patterns, sex drive, and the immune system. According to *Prime* magazine, melatonin may also help reduce the risk of heart disease, prostate cancer, and breast cancer.

Meditation Exercises

Breath-Counting Meditation

Using your free-word, close your eyes and "observe" the in-and-out flow of your breath. Free-breathing in, perform a mental count of one. After comfortably holding your breath for a count of three, slowly empty your lungs, mentally counting two. Repeat the inhalation, count three, hold, and exhale with the count of four. If you lose count, begin again with an inhalation at one.

This is an excellent quick relaxation-meditation technique to use when on a stressful night stakeout or while waiting in ambush. Even five minutes of this meditation has a calming effect.

No-Mind Meditation

Use your free-word technique to relax. Now, clear your mind of distractions by mentally repeating the phrase "no mind." Whenever your mind wanders, draw your mind back by again repeating "no-mind." With determined practice, this no-mind meditation becomes increasingly easy. Ultimately, no-mind will become a mantra chant (like free) that will help you clear your mind and relax with a single recitation of the phrase.

Ninja often practiced *misogi*, the purifying of the body and mind with cold. One misogi exercise required the ninja to meditate and chant a mantra while under an icy waterfall. The effects on the body from this and similar misogi practices closely approximate the effects of stress on mind and body, helping desensitize ninja to stress. Ninja also often meditated in total darkness for days prior to a night mission.

THE POWER OF VISUALIZATION

Our eyes are not cameras perfectly photographing everything they see and then storing a flawless copy in the brain. Instead, when we look at an object each of our eyes registers a different image. The brain then fuses these two images into one three-dimensional picture. So, when we look at a tree, we are not seeing the tree but rather our mind's re-creation of that tree.

Between the breaking down of an image and its reassembly inside our heads, there are dozens of different filters, including physical defects in the structure of the eye (e.g., color blindness), chemical imbalances in the brain (either natural or self-induced), and even our emotional beliefs and fears, all of which can affect how accurately the image is reassembled inside our head.

All humans use visualization in their daily lives. But whereas most people are unaware of doing so, knights of darkness purposely cultivate visualization as both a tool and a weapon. Research conducted at Stanford University proved that visualizing an image of action—seeing it in your mind's eye—fires the nervous system in the same way as actually performing the action being visualized. Such visualization is behind the success of many world-class athletes. In fact, 90 percent of sports figures practice some form of mental training. For example, shadow boxing employs visualization, as do the katas (ritual forms of practice) of karate. During martial arts kata practice, students first learn the overall pattern of movement—where to step, how and when to throw punches and kicks, and so on. But in order to master the kata, students must "see" (i.e., visualize,) an enemy standing in the way of their kicks and punches. Without this kind of visualization designed to train the nervous system to react in a real-life confrontation, katas are merely ballet.

Some experts argue that visualizing an action is superior to actual physical practice. According to Dachman and Lyons in their *You Can Relieve Pain: How Guided Imagery Can Help You Reduce Pain or Eliminate It Altogether:* "Effective imagery can have as significant an impact as actual physical experience. In fact, it can be argued that imagery can even have a greater impact because there is no external interference to the imaging process."[1]

Visualization Exercises

Afterimaging

Sit across the table from a partner who has placed at least nine small objects on the table. Fix the objects on the table in your mind and then close your eyes. Attempt to retain the afterimage of the objects in your mind's eye. With your eyes still closed, have your partner either remove one of the objects on the table or add another object to the collection. As soon as he has completed this task, open your eyes and immediately tell him which object has been removed or added.

SWAT teams use afterimaging to "flash" a room, which is peering around doorway and quickly pulling back, having "fixed" the contents of a room in their mind's eye. Knights of darkness fix the position of obstacles and occupants of a room in their mind's eye before plunging the room into darkness to facilitate escape.

Yantra

To develop their powers of visualization, ninja borrowed a training device from their thuggee cousins. The *yantra*, in its simplest form, is a dark symbol (a circle, square, and triangle) outlined on a light background.

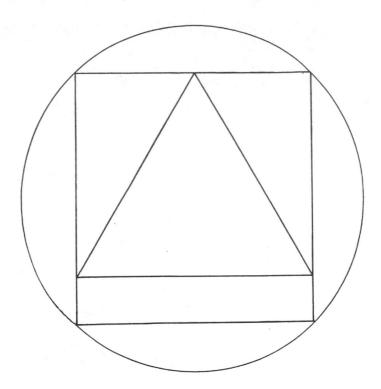

Illustration 1. Yantra visualization practice symbol.

Concentrate on the dark symbol for several minutes. Now close your eyes and keep the afterimage of the symbol in your mind's eye as long as possible. For the first few seconds this is relatively easy, because the dark symbol on light background has been "burned into" your retina. When the afterimage fades, begin the exercise anew. Each time you will be able to keep the afterimage in your mind's eye for longer periods of time allowing you to eventually graduate to increasingly complex yantra practice images.

ENDNOTE

1. Ken Dachman and John Lyons, *You Can Relieve Pain: How Guided Imagery Can Help You Reduce Pain or Eliminate It Altogether*, (New York: Harper & Row, 1990).

CHAPTER SIX

Training the Six Senses

The night's deceptions played tricks with my eyes. I could see inky black shapes moving against a background of the same inky black blindness.

—David Lovejoy
"Night Ordeal," *Gallery*

Moving through the night from shadow to shadow, the darkness is our cloak of safety. Wrapped in the blackness, we remain safe only through the full use of our six senses, each of which have their part to play in our successful and silent night movement. We must learn to use our senses not only to disguise our position and movement, but also to help us discern the presence and position of our enemy.

None of our senses stand alone; each augments the others. However, each sense is also unique, with its special part to play in our survival. It is important that we understand and learn the full use of these senses that, during the day, we more often than not take for granted.

DEVELOPING YOUR NIGHT EYES

Fully functioning human eyes can see a candle flame from 30 miles away on a clear, dark night. Overall, males are more sensitive to bright light and can detect more subtle differences in light than females. We use our eyes more than our other five senses combined. Unfortunately, our eyes depend on light to function, and light is limited at night.

Animals, have a sensitivity to light that man does not possess. Cats and other nocturnal animals have mirrorlike surfaces that repeatedly ricochet light once it enters their eyeballs similar to the way multiple aligned mirrors reflect a single beam of light between them several times. This reflective surface can amplify light entering a cat's eyes by as much as 50 percent, increasing the animal's chances of seeing an image at night by that same percentage. At first glance it might sound good to have eyes like a cat, which can literally see in the dark. However, an animal's increased sensitivity to light, resulting in acute night vision, comes at the expense of visual clarity. So, while human beings do not see as much as animals at night, what humans do see they see more clearly than do nocturnal animals.

Like all predators, man has eyes set well forward in his head. This allows him to better spot and focus on prey. When the human eye, or animal eye for that matter, sees a moving object, it first sees the movement of that object. The eye then makes out the object's silhouette. Finally, the eye discerns color.

To understand how knights of darkness use each of these—movement, silhouette, and color—to their advantage, we must first study how the normal human eye processes information. Once we understand how the eye gets its information, it is a simple matter to confuse an enemy's eye into misinterpreting incoming information.

The importance of understanding how the different parts of the eye function is comparable to, and just as important as, knowing how to field-strip a weapon in pitch black.

The Retina

The retina is the sensory membrane that lines the eye and acts like a movie screen, receiving the image formed by light as it passes through the lens. It is connected to the brain by the optic nerve behind the eyeball. Light captured by the retina is translated into electrical impulses that tell the brain what it is seeing.

The word *retina* comes from the Lithuanian *retis,* which appropriately means *sieve,* which is basically what the retina does: it sifts light coming into the eyes, encodes it, and then sends this encoded data along to the brain for decoding. In effect, the retina acts as the brain's visual gatekeeper.

To do its job, the retina relies on two kinds of optical receptors: cones and rods.

Cones are the receptors in the retina responsible for detecting color and discerning fine detail. They dominate during day vision. Cones respond most strongly to the primary colors, are most sensitive to green and yellow, but can nevertheless distinguish 300,000 color variations. Cones live off light; without light to stimulate them, they cease to function.

Rods, on the other hand, are the receptor cells on the retina responsible for night vision and peripheral vision; both are vital tools. Rods consist of millions of tentaclelike receptors designed to gather in light. There are more than 120 million of these rods per eye, all containing light-sensitive pigments that produce night vision and peripheral vision. Rods are not designed to assimilate colors. They are more comfortable with the hazy gray images in fading twilight and are most sensitive to colors in the blue and green range of the color spectrum. As a result, at night blue and green lights are much more visible to the human eye than red lights. This is why many emergency vehicles have switched from red to blue flashing lights. Red objects can actually appear black at night because of the rods' insensitivity to red.

Adapting to the Dark

This includes dilation of the pupil and an increase in retinal sensitivity. The retina's sensitivity to darkness depends on the existing amount of rhodopsin, a pigment found only in rods. Rhodopsin is more commonly known as "visual purple." The adjusting of the eyes to reduced illumination is dependent on the regeneration of visual purple.

So we know that light stimulates cones. Unfortunately, this same light also destroys or bleaches out rhodopsin in the rods. Conversely, darkness permits rhodopsin to regenerate.

Introduced to darkness, the light-sensitive eye increases its sensitivity at a rapid rate. In the first minute the eye is exposed to darkness, the retina becomes 10 times more sensitive to light as its supply of rhodopsin, bleached by the light of day, begins to regenerate. Within 20 minutes the eye has become 6,000 times more sensitive to light. After 40 minutes the eye has reached its maximum limit of light sensitivity. Fully dark-adapted, your retina is sensitive to 30,000 times less light, and you have attained your night eyes.

Ninja often required their night warriors to meditate in complete darkness for 24 hours prior to embarking on a night mission. This practice undoubtedly had its use in mentally preparing the ninja for the upcoming mission, since only 40 minutes in full darkness would have been enough to establish a ninja's night eyes.

Using Visual Purple as a Weapon

Since daylight bleaches visual purple, any bright light at night also destroys night vision. Likewise, any sudden flare of light (a "flash-bang" grenade, camera flash, or a lighted match) will temporarily stun your night vision, blinding you. A sentry guarding the entrance to an installation is temporarily blinded each time a vehicle flashes its headlights in the direction of the gate. Understanding how visual purple works allows someone to time his penetration of an installation, slipping past the sentry while the sentry is temporarily night blind. Therefore, when on guard duty it is important to look outward, away from a lighted perimeter or campfire instead of inward toward the light.

Consequently, night fighters strike either while the sentry is facing the light or as soon as he turns back to face the dark, before his visual purple has had a chance to regenerate. The oldest trick in the book—distracting an enemy by throwing a rock over his shoulder—can be used to make a sentry turn in the direction of the brightest light, thus destroying his night vision. Strike as the sentry turns back toward the dark, before his eyes have had a chance to adjust from light to darkness.

In his best-seller, *Papillon*, ex-convict turned author Henri Charriere describes the ploys inmates used to kill their fellow inmates on Devil's Island and other prisons in the French prison archipelago off the South American coast in the first half of the 20th century. One such violent incident described by Charriere involved an inmate's killing two fellow inmates at night. The killer accomplished this by approaching the beds of his intended victims in the middle of the night and flashing the bright beam of a flashlight directly into their eyes. Blinded, both men were quickly dispatched.

Peripheral Vision

Peripheral vision is the ability to perceive movement at the edges of our vision field, where we see only in black and white. This is because our peripheral vision is controlled by our color-blind rods. These sensitive night receptors are more concerned with drawing in all available light, color being secondary in importance to discerning movement and silhouette. This is why we see movement "out of the corner of our eye" without being able to identify what it is.

Any decrease in the level of light helps increase our peripheral vision. Since rods give us our night vision, any decrease in light level, natural or artificial, helps activate our night vision.

Improving your awareness of your peripheral vision, for both day and night use, is a vital skill in the arsenal of accomplished night fighters, and the best way to consciously activate your peripheral vision is to practice *defocusing*. Pick a spot on the horizon directly ahead of you, such as a ridge in the countryside or a crosswalk on a crowded city street. Focus your total attention on the center of this picture. Now slowly narrow your eyes. Narrowing your eyes decreases the light coming into your eyes and helps reactivate your visual purple, especially around the edges where your peripheral vision receptors are located. Narrowing your eyes to slits, notice the colors beginning to dull, which is a direct result of increased rod activity. Keeping your eyes narrowed, loosen your intense focus on the center of the scene, allowing your vision to blur slightly. As you defocus, you will begin to notice more movement out of the corners of your eyes. You will not be immediately able to discern exactly what it is that has caught the attention of your peripheral vision, but you will be able to recognize that something is moving at the edges of your vision. You can then focus on the particular disturbance to determine if it is friend or foe.

Regularly practicing this defocusing technique increases your overall awareness of your peripheral vision. The technique is similar to the meditation technique known as *zazen*. In zazen, students sit in concentration with eyes half-closed and defocused. As the meditation progresses, colors slowly fade from the spot being viewed as the half-closed eyes begin activating visual purple rods. The fact that the meditator is relaxed improves his peripheral vision. Any type of tension decreases the efficiency of the senses.

Increasing Your Night Vision

Ancient astronomers didn't have ground-glass lenses to magnify their eyesight when staring at the night sky. They had to rely on simple, straight tubes. Although such tubes didn't magnify the object being viewed, per se, these crude telescopes did cut down on the available light coming into the astronomer's eye. As a result, rods were reactivated and visual purple regenerated in the astronomer's eye. The small amount of light seen through the telescope was then used to maximum efficiency. And since rods also control peripheral vision, astronomers found they could see stars better by not looking directly at them.

If you cannot afford an expensive electronic night scope, then focus and increase your night vision by making a viewing tube with your hand.

The Enemies of Night Vision

Any flash of bright light, no matter how temporary, will

Illustration 2. Increasing your night vision.

destroy night vision. It then takes time (seconds to minutes, depending on the intensity of the flash) for your visual purple to regenerate.

Bright flashes of light can either work for or against you. For example, flashes from lightning can destroy your visual purple as easily as they destroy the visual purple of an enemy sentry. Flames from cigarette lighters and matches also destroy visual purple. When stalking a sentry who you know is a smoker, simply wait until he lights a cigarette (with a regular smoker, every 15–20 minutes) and then make your move as soon as the flash has ruined his night vision.

Flash-bang grenades and the camera flashbulbs can both rob sentries of their night vision, although the combat applicability of these must be judged on a situational basis. Remember how vehicle lights affect a sentry's night vision? If you cannot wait for a vehicle to approach a targeted compound, have a confederate, preferably a good-looking woman when dealing with male sentries, pull her car up to the sentry box and ask for directions. Use the blinding light of her headlights to slip past the sentry. Bivouacked soldiers looking too long into a campfire or facing bright compound lights will also be night blind.

To approach a sentry close enough to disable him, walk directly up to him, pretending to be a superior officer, all the while shining a flashlight directly in his eyes. This will destroy his night vision, and you will appear only as an approaching silhouette, one he will still be trying to identify when you cut him down. To augment this ploy, disguise your silhouette by wearing the appropriate cap or helmet.

Always beware, lest your light inadvertently alert others. Whenever a driver parked in an unlighted area opens a car door, the overhead light comes on, temporarily night-blinding him. When traveling by auto, modern knights of darkness disable interior car lights by using the handyman's secret weapon: duct tape.

You must guard your night vision when moving in or near any structure where lights might be suddenly turned on. When at all possible, any time you enter a room on a night mission, disable the lights (lamps, ceiling lights). Whatever activity you need to carry on inside a structure, a simple penlight should be sufficient for illumination. Any time you suspect that a bright light might be used to blind you, close one eye. Use your open eye to see, while keeping your closed eye in reserve. Immediately after you have been flashed by a bright light, open your reserve eye. This allows you to see while the first eye regenerates its visual purple.

Man inherited a "freeze reaction" from his animal ancestors. Many incorrectly believe that an animal that freezes when the beams of headlights catch them in the middle of the road does so out of fear. This is not the case. While it is true that bright headlights rob the animal of its night vision, an animal freezes instinctively because most predators rely on movement and sound to pinpoint prey. The human body also freezes temporarily when blasted with light. This is a carryover from a time long ago when we were prey rather than predator. This reaction also occurs because, blinded, we might accidentally stumble into danger. Animals freezing in headlight beams become roadkill. Likewise, night fighters freezing when "ambushed" by a bright light also become history.

Chiaroscuro

Chiaroscuro is defined as the ever-shifting play of light and shadow. You must thoroughly familiarize yourself with the nuances of chiaroscuro to better cloak yourself and your mission in darkness.

The brighter the light, the darker the shadows. Two factors complicating this simple equation are multiple light sources and shifting light sources.

Double light sources can either reinforce or cancel out each other's shadows, with the brighter and closer light producing the dominant shadow. For the night warrior, it is important to be constantly and consistently aware of shadows created by moonlight, passing vehicles, and perimeter and structure lighting. You must map out the safest path leading to and from the objective, a path that takes full advantage of chiaroscuro. You must also correctly predict where shadows will appear and disappear throughout the night. In other words, you must be able to anticipate shadows that will be created or destroyed by such things as the moon, or the turning on and off of perimeter and structure lights. You must also be aware of how shadows within a structure differ from those outside a structure.

Shadows within a structure depend upon the positioning of lights in the room (e.g., table lamps, overhead lighting) as well as any mirrors and other reflective surfaces within the room. Except for light coming through windows, most interior lighting is fixed in place.

While some lighting (and their shadows) outside a building may be fixed (e.g., porch lights, perimeter fence lights), other shadows outside a structure are constantly shifting, because they are produced or influenced by the passing of the sun or moon, cloud cover, vehicles, and flashlight-carrying sentries. Like sunlight-produced shadows, moonlight-produced shadows change hour to hour as the moon travels across the night sky. Moonlight-produced shadows can also change minute by minute depending on the amount and speed of drifting cloud cover. The length and placement of shadows vary as the day or night progresses. The shadows you use at 9 P.M. entering an area will not be there for you when you exit six hours later.

In anticipation of movement through an unfamiliar area, or when planning to infiltrate an installation or enemy bivouac, you must first perform an in-depth reconnaissance of the area of operation (AO) to determine the most expedient route that offers the best cover. A vital part of this recon is determining where the deepest shadows are and where they will be during your movement through the area. This requires learning to calculate where the deepest shadows will be as hours progress.

When moving up valleys or down paths with high brush or timber to one side, you want to stick to the side of the valley that is most deeply shadowed. However, depending upon how long your mission is scheduled to take, and whether you plan to return along the same route—seldom a good idea—the shadows that hid you while you were moving up the valley will not be there as you move back down the valley.

Any night movement is an exercise in patience. Wait for shadow cover and go out of your way, literally, by taking a roundabout route if necessary to gain the advantage of deep shadow cover.

Many of the same rules for moving through the countryside apply when moving through an urban setting—between buildings or down an alley—where you will also find shadows chang-

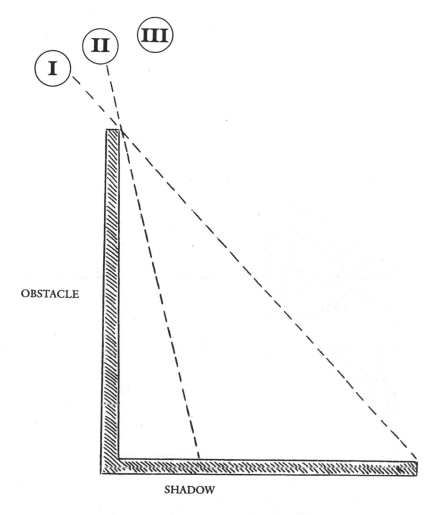

Illustration 3. Sunlight- and moonlight-produced shadows.

I: The shadow produced by the rising of the moon, when it is still below the lip of an obstacle, will be deep and dark. As the moon climbs, the shadow shortens and lightens as more moonlight fills the night sky.

II: As the moon rises higher in the sky, over the lip of the obstacle, the shadow available decreases and the shadow you used an hour before will not now be there to offer the same concealment.

III: As the moon reaches its zenith, many shadows disappear. As the night progresses toward morning, shadows will appear on the opposite sides of the obstacles, necessitating rerouting your egress.

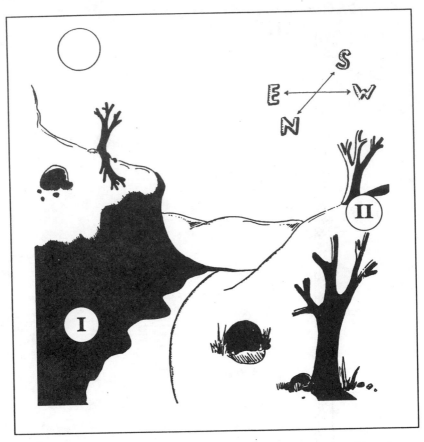

Illustration 4. Movement of moonlight shadows in countryside.

I: Moving south through a valley, the moon, having risen in the east, casts the east side of the valley into shadow. Once the moon reaches its zenith, the hollow of the valley will be illuminated and most shadows will be gone, except for those produced by overhang. As the night progresses, shadows return on the opposite (west) side of the valley.

II: When large shadows are unavailable, use smaller shadows produced by trees, boulders, ditches, and defiles to shield your movement.

KNIGHTS OF DARKNESS

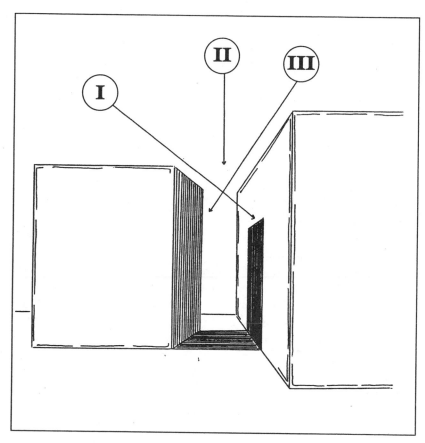

Illustration 5. Movement of moonlight in urban settings.

I: The moon in back of a building creates a shadow. This shadow first shortens and then lengthens as the night moves toward morning. A building casts one shadow on an adjacent building at 9 P.M. and quite a different shadow at 3 A.M. Note how the building at 9 A.M. creates a three-sided shadow—on itself, on the adjacent building, and on the intervening ground between the two buildings.

II: At midnight, these shadows cease to exist as the alleyway is lighted by the moon directly overhead.

III: By 2 A.M. the shadows will be on the opposite side. Additionally, the fact that the building on the right is taller means that this building will cast an additional shadow on the roof of an adjacent shorter building on the left. Shadowed roofs can be useful when adjacent shorter buildings create the possibility of observers in the taller buildings seeing you as you move across the moonlit rooftop.

ing as day or night progresses. Shadows in an urban setting are influenced not only by the movement of the sun and the moon but also by such factors as weather, reflective surfaces, moving lights from passing vehicles, blinking neon, and the turning on and off of interior and exterior structure lights.

Once you have mastered anticipating the lengthening and shortening of shadows, as determined by light sources and the progression of the night, you can begin to use such knowledge as your cloak to help you move toward your goal more safely.

The most fascinating time for the play of chiaroscuro shadows is twilight; either right before the sun goes down or just prior to its rise. At twilight the sky is light and the ground is dark, and human eyes have a difficult time adjusting from the bright sky to the dark ground.

Nature, in her wisdom, realized early on the importance of coloring her creatures to survive, especially at twilight, when many of them emerge from hiding. Most animals, rabbits for example, have darker fur on the top of their body and lighter fur on their bellies. This darker top coloring helps hide them against predators from above. As their camouflage mechanism, raptors are lighter on their bellies so they can't be seen from below. When a ground dweller, a rabbit for example, looks up into the sky, a circling hawk is practically invisible against the light sky. Likewise, to the circling predator looking toward the darker ground, the rabbit, with its darker fur on top, is difficult to see. This type of color camouflage is one reason predators hunt primarily through spotting movement.

Sentry towers placed high above the ground are most vulnerable at twilight, since they receive light from the setting or rising sun while the ground around them is still shrouded in darkness. This makes it difficult for sentries to adjust their eyes from the bright of the sky to the dark of the ground. Ground-level sentries do not have to worry about the setting sun's interfering with their night vision, which develops naturally with the coming of darkness. As previously discussed, however, ground-level sentries are susceptible to vehicle lights and moving shadows.

Most fixed security perimeters prefer aboveground boxes

Illustration 6. The effects of twilight.

I: The sentry in the tower is "twilight blind," making him easier to slip past.

II: Ground-level sentries are better equipped to handle twilight, provided they do not allow their eyes to linger on the setting or rising sun.

as they offer a better panorama as well as increased security for sentries.

Silhouette can be both a liability and a weapon. Casting a silhouette of your body against any lighter surface must be avoided. More often than not, when a sentry spots your movement, what he will see first is the movement of your shadow or silhouette rather than your actual person. Even in dim light, your silhouette can show up against a lighter background, such as a wall. Security-conscious installations paint the exterior walls of the first floor of buildings a lighter color than upper floors to make it more difficult for a nocturnal intruder to move along the walls.

Always avoid skylines, whether in the countryside or when perched high on a rock ledge. Never stand in front of windows or doors; someone might suddenly turn on a light inside and silhouette you against the window or "frame" you in the doorway. The rule for all night fighters is to pass all windows and doors as if they were "active," that is, as if light might suddenly flood out from them.

When moving down hallways, alleys, and similar approaches, always glue yourself to one side or the other so as not to be silhouetted in the center; by staying close to a wall, you present less of a target.

The silhouette of your body is normally distinctive and can identify you as an intruder. Seen in silhouette, an enemy trooper can be identified through the distinctive silhouette of his uniform (cap, helmet, straight or baggy pants) and by the outline of his weapon (M16 vs. AK-47, for example). This is one reason why Spetsnaz and other infiltrators carry weapons similar to or the same as those of their enemy.

The night fighter can disguise his silhouette so that it appears he is not carrying a weapon. This allows him to get closer to an unsuspecting sentry or target. When crossing a street in a city or when walking through a shadowed compound, the night fighter keeps his weapon flush with his leg, disguising the fact that he is armed. In those times when you are caught without a weapon, place an object in your hand to make your silhouette appear as if you are armed. At the very least, such ploys place doubt in an enemy's mind, making him hesitate long enough for you to approach him and get within striking distance. (We often hear stories of police shooting an armed burglar, only to discover, when turning on the lights, that the silhouette they saw "brandishing a weapon" was just a nosy neighbor carrying a drill or a golf club; in silhouette, the drill was mistaken for a handgun, the golf club for a rifle.)

Medieval ninja were masters of disguising their silhouette. One often overlooked weapon was the cloak, which was used to cover the ninja and make him appear in silhouette to be a boulder, tree, large vase, or box. This technique allowed ninja to throw off pursuers or wait patiently in ambush for the arrival of an unsuspecting victim. The technique contributed to the widespread belief that ninja were shape-shifters who were able to

Illustration 7. Disguising your silhouette.

I: In the night fighter's true silhouette, he can easily be seen to be carrying a rifle and wearing a sidearm. His military helmet and pants bloused into combat boots can also be easily identified.

II: Disguising his silhouette, the night fighter places the rifle behind or flush with his leg, drapes his hand in front of his sidearm, removes his helmet, and unblouses his pants.

assume any form at will. You can use such ploys to remold your silhouette to fit your surroundings. To elude pursuers, you can adapt this principle and appear as a tree stump, shrub, signpost, and so on. You can also make objects (lampposts, signs, bushes) appear to be human by placing a hat or a shirt with an upturned collar on them, which can distract an enemy long enough for you to get within range.

A popular technique that employs silhouette as a weapon is to replace an enemy sentry with a member of your penetration team, a simple matter of having one of your team don the distinctive headgear of the removed sentry and shoulder the sentry's discarded weapon. With your team member standing in place of the sentry, his silhouette will pass for that of the fallen sentry, provided your substitute sentry doesn't expose himself to the light or close scrutiny.

When in a house alone, a woman can place a man's hat on her head and wear a heavy jacket (disguising her breasts) and then walk in front of a window shade. To any stalker, her silhouette will appear to be that of a man.

"Moving" shadows cast by blinking lights or passing vehicles, as well as those produced by clouds drifting overhead, can also be useful. With blinking lights, such as neon signs, you can time the length of darkness and move through the shadow before the sign comes back on. (Blinking lights interfere with an observer's night vision, because looking in the direction of the blinking lights causes his visual purple to be "bleached out" each time the light comes back on.)

In the case of drifting clouds, whether in a rural or an urban setting, use the same technique by timing the movement of clouds to stay in their shadows.

For the night fighter, the trick is the masterful use of the eye while not limiting one's "seeing" to the use of the eye only.

HEARING

Few animals have hearing as poor as man's, although many animals have poor eyesight. They compensate for this by using

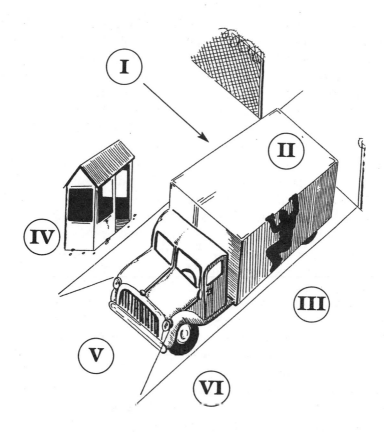

Illustration 8. Using moving shadows.

I: Moonlight or artificial light on one side of a vehicle casts the opposite side of the vehicle into shadow.

II: When a vehicle's top is above the light level of artificial lighting, the top of that vehicle will also be shrouded in shadow.

III: The night fighter can attach himself to the side or roof of a large vehicle (truck, railroad car) to take advantage of the shadows on the vehicle in order to pass into or out of a closely guarded perimeter. Other times, night fighters can move in a crouch-walk, hidden by the moving shadow cast by a passing vehicle.

IV: Light from other artificial sources (sentry boxes, passing vehicles, perimeter lighting) impacts on the play of shadows.

V: Vehicle lights can destroy shadows and be reflected back from shiny surfaces within the perimeter.

VI: Always beware of mirrors and shiny windows on vehicles, which might reveal you to the driver. Also be aware of interior cab lights that will illuminate an area when the driver opens his door.

their other senses more fully than does man. For example, owls have eyes that are extremely sensitive to dim light, but they do not rely on their eyes to the exclusion of their other senses. As a result, an owl can catch any moving creature, even in pitch dark, by sound alone. To make up for man's barely adequate hearing, night fighters can use what they do have more effectively.

Our ears pick up sound from vibrations impacting the delicate bones of our inner ear, and sound vibrations often act as hitchhikers, riding on water and wind much farther than they normally would on a still day far from water. As with scent, sound travels farther in the direction the wind is blowing. This is why it is vital for the night fighter to approach a target from downwind. When moving over or through water, the night fighter must beware of stirring the water unnecessarily, since these vibrations can be carried across water or downstream to alert an enemy.

Vibrations can be "heard" through metal, wood, and even the ground. American Indian trackers placed their ears to the rails to determine the approach of the "iron horse," or to the ground to hear the approach of horses or bison. Today, heavy-footed troops or rumbling vehicles can also be heard by feeling the vibrations in firmly packed earth. This is one of the advantages barefooted indigenous guerrillas have over boot-wearing invaders. This type of hearing—discerning vibrations through feel—can also be placed under the heading of "touch," since it is often a case of "hearing" with your hand, bare feet, or some other part of your body.

When listening to rails or some other metal (copper pipes or vents in a building, for example), you are not receiving the vibrations with the bones of your ears as much as you are hearing the vibrations impacting against your skull as a whole. Likewise, when feeling ground vibrations, you are simply using your hands or feet as sound receptors in place of your ear.

In the absence of expensive directional microphones and other listening devices, improve your hearing by cupping your palms behind your ears. Close your eyes, open your mouth to a wide "O," and flare your nostrils with your nose slightly up, as if sniffing the air. Tilt your head in the direction you are listen-

Illustration 9. Improving your hearing.

ing. This position helps you take full advantage of your cranial cavities to collect sound vibrations. Relaxation is the key. Scan the area and slowly turn your head first one way and then the other in a 180-degree arc.

Whether in the bush or in an urban setting, you must be familiar with the normal sounds of your AO. What you don't hear is as important as what you do hear. If you are moving through a wooded area, listen for animal sounds; you may hear startled animals, indicating that an enemy is approaching. On the other hand, you may suddenly "hear" silence, which is also indicative of the approach of danger.

Having familiarized yourself with the normal sounds of your AO, you should be able to immediately discern any out-of-place sounds, such as normally noisy woodlands suddenly becoming quiet and still or the sound of an unfamiliar engine pulling into your driveway. American Indians, ninja, and other accomplished night fighters used animal sounds and natural noises, such as that of a bush being rustled by the "wind," to communicate surreptitiously.

Ninja often took trained or drugged animals with them on night missions. Whenever a ninja thought a samurai sentry heard something, the ninja released an animal (a fox, for example) to make the sentry think it was only an animal that he had heard. (In Japanese mythology, foxes are bad luck, analogous to the Western superstition of black cats. Upon seeing a fox, a superstitious samurai sentry was twice as likely to back off from any further investigation. This kind of animal ploy can still be used today to play havoc with modern perimeter motion sensors.)

As ninja are fond of reminding us, *there are no new answers, only new questions.*

If you have ever heard something or had a feeling that you were being watched, you were probably right! Millions of years of the survival of the fittest can't be wrong. Therefore, always err on the side of safety, both yours and your teammates. The blood that floods your face when embarrassed by making a mistake or being paranoid passes. The blood from a cut throat doesn't.

SCENT

Our neglected sense of smell is a good indication of just how far man has strayed from his place in nature.

Although women have a better sense of smell than men, for the most part we ignore our sense of smell and rely more on our eyes for distinguishing friend from foe. We then compound the problem by perfuming our lives with overpowering deodorants and pine-scented cleaners that mask any natural warning scents that reach our nose.

In the animal kingdom, predator and prey alike rely heavily

on their sense of smell to survive. Potential prey depend on their sense of smell to warn them of the approach of predators. For this reason, predators instinctively stalk their prey from downwind, with eyesight taking over only after the prey has actually been spotted.

Pheromones, which are chemicals that we discern and then decipher as smell, odor, or aroma, cling to almost every surface and are carried on wind and water in the same way as sound vibrations. They can cling to surfaces for long periods of time, which allows dogs to track people merely by getting a whiff of clothing that the person being tracked had worn.

Dogs rely much more on their sense of smell to distinguish friend from foe than they do eyesight. Professional burglars entering a house where they suspect a dog may be, first rub their hand over the doorknob of the house to pick up traces of pheromones left by the owners of the house. When the family dog approaches, the burglar stretches forth his hand, with the familiar smell, for the dog to sniff. This ploy will placate many untrained family dogs. A variation of this ploy involves actually obtaining a piece of clothing, such as a jacket, belonging to the person whose house you are invading.

Nineteenth century Arab assassins infiltrating Foreign Legion posts did so naked, with the belief that dogs will not bark at a naked man.[1] There may be some truth behind this belief, because without their pheromone-saturated robes to identify them, these Arab sappers would have given off less of a scent "signature," therefore temporarily confusing the guard dogs.

Much has been made of man's "peripheral body sense," which is his ability to sense when someone is within three to five feet of him. Some knights of darkness dismiss the idea that this sense is a sign of ESP, instead maintaining that it is merely our ability to discern smells at a subconscious level.

Another part of being familiar with your AO is being aware of its natural odors, so as to be able to better discern any out-of-place smells that might warn you of danger. Out in the bush, heavy deodorants, flatulence, tobacco, gunpowder, and weapon oil can all give you away, just as it can warn you of others in the area. During the Vietnam War, U.S. Special Forces and Ranger long-range recon patrols (LRRPs) dieted on rice and other indigenous foods prior to going deep in-country. This precaution was necessary because the Vietcong, intimate with the smells of his jungle, could smell the difference between the sweat and flatulence of meat-eating Americans and that of rice- and fish-eaters, like most Vietnamese. Any latrine leftovers smelling of undigested meat were a dead giveaway that Americans were in the area.

American Indian hunters smeared themselves with the scent of the animal they were stalking, such as bison. This is also the reason why such night fighters as the wolfshirts and the leopard cult killers cloaked themselves in the skins of animals. For one thing, such skins would confuse guard dogs, perhaps even frightening them off. Second, these skin cloaks helped mask the body's natural smell. Artificial animal smell ("bitch scent") can be used today to distract and confuse guard dogs when you are penetrating a perimeter or eluding pursuers who are equipped with bloodhounds.

TOUCH

Our skin is sensitive to hot and cold, subtle changes in the wind and barometric pressure, and vibrations from a myriad of sources. The skin on our fingertips and face can perceive pressure that depresses our skin a mere .00004 of an inch. Whether on assignment in an urban environment or hoofin' it through the boonies, night fighters use their mastery of their sense of touch to warn them of the approach of danger and the presence of enemies and to guard their own passage.

When expecting to fight foes in complete darkness, teams of ninja would strip down to their loincloths or go naked before infiltrating. That way, anyone they touched in the dark who was wearing clothing was the enemy.

Vibrations
Vibrations in the ground—whether through railroad iron or

other metals, the shaking of wooden stairs, vibrations on a wooden-slat floor, or water lapping against you as you wade through what should be tranquil water—can all warn of the approach of potential danger. You must remain aware of your movement, lest you inadvertently create such vibrations.

Modern motion detectors and ground listening devices, like many devices and machines initially designed to extend our senses, often end up dulling and replacing our natural senses (in this case, our sense of touch). Despite technological developments, nothing is better than developing your own innate senses and abilities.

Heat Detection

Heat detection is another way you can use touch as both a defensive tool and an offensive weapon. Inside a house, the heat felt on a chair or in a rumpled bed indicates that someone has recently occupied it. Heat felt coming off the hood of a vehicle indicates that it was recently running, with the driver perhaps still lurking nearby. Out in the bush, warm, pressed-down grass or a warm indentation in the ground indicates that someone or something has recently rested there. Casings of recently fired bullets will also be warm, even when fired from a silenced weapon, and indicate the recent presence of an enemy. You must, therefore, avoid resting for too long in any one spot while working your way through the bush, lest your body heat left on grass or the ground betray your presence or route.

TASTE

In ancient times, court food tasters were specially chosen or bred for their ability to discern minute amounts of poison in food. Human taste buds can discern .04 of an ounce of table salt dissolved in 530 quarts of water. Experienced sailors can taste the salt in the air when near the sea. There are night fighters who also cultivate their ability to taste the air to discern odors through their tongues or, more specifically, the soft membranes under the tongue. (The soft membranes under the tongue are sensitive and absorbent. Cigar connoisseurs swirl tobacco smoke under their tongues to allow nicotine to be absorbed through this soft tissue. It is therefore possible, with practice, to learn to taste the air to discern certain odors too subtle to be picked up by the nose.)

To cultivate your ability to taste the air, purse your lips as if about to whistle and slowly draw air in. Allow the air to swirl around and under your tongue. Try to discern salty from sweet from oily. With practice you will be able to discern perfume or deodorant (a sweet or oily taste under your tongue) and residual tobacco (sharp, pungent) left in the air or on clothing by a passing enemy.

Some argue that "tongue sniffers" are actually picking up odors through their noses, subtle odors that only register on a subconscious level. To night fighters the point is moot. If a tactic, technique, or training method works, use it, no matter what its origins. In other words, whatever gets you through the night.

THE SIXTH SENSE

Many people swear by the existence of a sixth sense, a form of ESP that accomplished night fighters develop to alert them to the approach of danger. Debates continue in the halls of academia over whether ESP actually exists. For example, we know that all humans have an electromagnetic field that extends out from their body and that we are often subconsciously, if not consciously, aware of someone stepping close to this field.

You can train to develop your peripheral body awareness to where you can sense the presence of another. As noted, some argue that what you actually sense are subtle smells (pheromones) or below-audible sounds that register at a subconscious level. Whichever explanation sounds the most plausible (perhaps it is a combination of both) is also moot. For us, in the end, all that matters is survival and the success of our mission.

Someone once estimated that human beings only use 10 percent of their abilities. More correctly, we use our abilities to only 10 percent efficiency. One need only look at documented accounts of East Indian yogis to realize humans do possess abilities that they do not use. So far as extrasensory perception is concerned, at the very least, by paying more attention to the development of your sixth sense, through diligent attention, concentration, and visualization, you will become more aware of your verified five senses. And, if nothing else, mastery of these five senses, used fully and in concert with one another, will make your enemies believe that you do possess a sixth sense and are, therefore, the most dangerous foe.

ENDNOTE

1. Dr. Haha Lung, *Assassin! The Deadly Art of the Cult of the Assassins*, (Boulder: Paladin Press, 1997).

CHAPTER SEVEN

Camouflage

The word *camouflage* comes from the French *camoufler*—to disguise—and therefore from the Parisian vernacular referring to smoke being blown into a person's eyes. For knights of darkness, camouflage involves a mixture of stealth, chiaroscuro, dress, and deportment. We employ camouflage to disrupt our movement and form. To be more accurate, what we actually disrupt through our use of camouflage is our enemy's perception.

The night fighter masters ways of disguising his position and direction of movement. To accomplish this, he distorts his form and silhouette so as to make the enemy unaware of his presence.

MOVEMENT DISRUPTION

The night fighter must master the ability to move without being seen. To accomplish this, he blends his knowledge of shadow and silhouette with specialized methods of stepping and approaching sentries that take full advantage of all available cover and concealment. To do this, the night fighter uses such objects as trees, shrubs, tall grass, rocks, and man-made structures to disguise or conceal his figure and movement. He also learns the proper way to move alongside walls and other structures so as to better blend into the backdrop and disguise his direction of movement.

FORM DISRUPTION

A bird spots a tiny moth and swoops in for the kill. Suddenly, where the tiny moth was a second ago, now appear the large eyes of a creature twice the size of the moth. For his own safety the diving bird veers off. What happened to the tiny moth? He merely spread his wings—wings decorated with large, colorful ovals that, to the bird at least, resemble the eyes of a larger creature. By disrupting its form, the moth survived another day. This same disruption of form is found in frilled lizards that inflate their neck shields to distort their form and make them a more formidable, less enticing target. In *Macbeth*, warriors cut pieces of Birnam Wood to disguise their form and their numbers as they attacked Macbeth's castle. During World War II, at the Battle of El Alamein in North Africa, Allied troops set up string-animated wooden cutouts called "Chinese soldiers" to draw enemy fire.

These are examples of disruptive camouflage, which uses false form or a distortion of shape to cloak the night fighter and confuse the enemy. Night fighters use two basic tools to distort their form and camouflage themselves: clothing camouflage and skin camouflage.

Clothing Camouflage
The cloaks of ninjas and wolfshirts are excellent examples of how clothing can be used by a warrior to distort his form. Modern camouflage clothing uses the same principle ninja's used (appearing as a shrub or some other object) to distort form by permitting the night fighter to blend into the background. This makes it difficult for anyone looking in his direction to spot where the background leaves off and the outline of the night fighter begins.

Camouflage clothing and camouflage skin paint are designed to manipulate color and to confuse the eye of an onlooker. Even though it is more difficult to see color at night, the lighter and

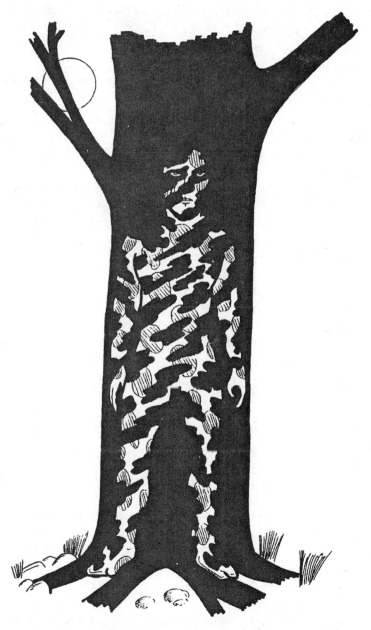

Illustration 10. Blending in with your background.

darker shades of camouflage still help to blend the night fighter into his background.

The eye first sees movement and then discerns silhouette. When otherwise properly camouflaged, if you remain absolutely still and take pains to distort your silhouette, there is little chance of an enemy spotting you.

Your clothing color should be appropriate to the locale and the background you are moving against. For example, wearing jungle camouflage clothing in a desert or on a snow-covered landscape would be suicidal.

An excellent tale emphasizing the use of color comes to us from the medieval *sDop-sDop* warrior-monks of Tibet. An evil king was persecuting Buddhists, and a bold sDop-sDop decided that the bad karma (repercussions) he (the monk) would incur from slaying the evil king would be less than all the bad karma the ignorant king was accumulating through his persecution of Buddhists. One day, while a festival was being held in the courtyard of the king's residence, a mysterious magician, dressed all in black and seated upon a shining black horse, trotted into the courtyard. Remaining astride his black horse, the dark-cloaked magician performed several acts of conjuring to the delight of the king, his courtiers, and guards. While the king and everyone else was still applauding, in one smooth movement the black-cowled figure drew a small bow and shot the evil king through the heart with a poisoned arrow. In the confusion that followed, the black rider escaped, but the king's bodyguard quickly mounted a chase.

The pursuers knew they were not far behind the black-garbed assassin when, after fording a deep river, they came upon a white-robed monk watering his white horse. The monk had seen the fleeing assassin and quickly pointed the pursuers in the right direction. Grateful, the posse thundered on. The white monk was, of course, the sDop-sDop, having reversed his robe (black on one side, white on the other) and having washed the black dye from his white horse as he forded the river just minutes ahead of the posse.

Modern-day night fighters often use reversible jackets and hats to throw off pursuers.

Skin Camouflage

Down through the centuries, various kinds of war paint have been used by knights of darkness. Such body paint serves the double purpose of frightening enemies and helping camouflage the wearer. During World War II, newly organized commando units experimented with various types of skin camouflage. In his *Deception in World War II* (Oxford University Press, 1979), Charles Cruickshank lists the results of several skin camouflage substitutes that can be used in case of emergency. His critique:

> **Cocoa:** satisfactory, but not waterproof.
> **Soot:** satisfactory, but not waterproof.
> **Mud:** dries too light.
> **Printer's ink:** too much shine, difficult to remove.
> **Cow dung:** didn't darken well enough, carried danger of tetanus.

Today there are several brands of reliable, waterproof camouflage skin paint commercially available. In addition, there are several brands of cosmetics (e.g., mudpacks, tanning gels) that might double as skin camouflage in an emergency. Experimentation, taking into consideration such factors as availability, skin irritation, and ease of removal, will yield several workable substitutes.

One further consideration: When operating in an urban environment, you might have to blend in with a crowd to elude pursuers, and you will find difficult-to-remove skin camouflage a liability rather than an asset. Where applicable, a camouflage face hood works just as well and, when necessary, is more easily discarded.

To recognize a head, we first try to discern the position of the eyes. Should only one eye be seen, we are at a loss as to whether the camouflaged person is to the left or the right of the single eye seen. This makes it very hard to target a camouflaged person, *if* he is spotted. This might seem like a trivial matter, but it can spell the difference between your getting shot right between the eyes or having a bullet whiz past the side of your head.

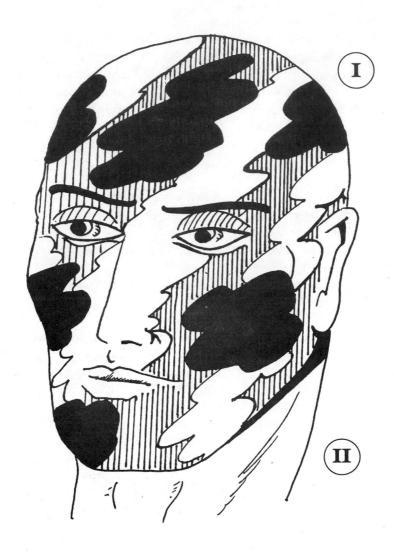

Illustration 11. Face-paint camouflage.

I: Always pass camouflage paint across the eyes so as to better disguise their position.
II: Make sure to cover the neck.

Cover all areas of skin that are reflective or that tend to be in motion—especially the eyelids, which are constantly in motion. It does no good to apply face paint to disguise your face when you have two white spots in the middle of your face that are blinking off and on. Also, be sure to cover the backs of your hands, since the hands will also be in motion much of the time. Care should be taken to disguise or otherwise cover fingernails and teeth, as their surfaces are very reflective.

Be sure to camouflage any weapons or other equipment to reduce their chances of being spotted because of distinctive outlines or reflective surfaces. Remain especially aware of potentially reflective surfaces, such as knife blades, ammo cartridges, scopes, and eyeglasses.

CHAPTER EIGHT

Methods of Movement

Having mastered the training of your senses and having acquired a good grasp of the use of camouflage, you are now ready to learn how to move from place to place at night without giving yourself or your teammates away. Many factors impact on how and when you move as well as why you might choose to move in one manner as opposed to another.

In general, we divide methods of movement into two areas: urban movement and rural ("bush"). Whereas several principles of night movement remain constant regardless of locale, there are enough unique differences between urban and bush movement to warrant our examining both.

URBAN MOVEMENT

Urban movement provides you with obstacles and challenges that you are not likely to find humpin' through the bush. Shadows and silhouettes are different in the city than they are in the countryside, in great part due to artificial lighting.

In the city, shadows are created by tall buildings and destroyed by passing vehicles. Shadows will be deep in alleyways where light from the street fails to reach. City streets may be brightly lighted and adjacent roofs above the "light line" (discussed in a moment) may prove to be the best shadowed path down a street.

It is important to understand shadow and silhouette in such a setting, since enemies may use their knowledge of shadow and silhouette against you. In the city, an enemy may tail you, moving from shadow to shadow or moving closer to you without raising your guard, simply by disguising his silhouette.

You need to master silhouette to disguise your weapons when forced to openly cross a street, a playground, or a backyard (see Illustration 7). When moving along darkened sidewalks, take full advantage of the deep shadows by walking on the side of the street in moonshadow while avoiding the lights cast by storefronts and passing vehicles. Use deeply shadowed doorways to help you elude pursuers. When you step into a heavily shadowed doorway, anyone following you or watching you from across the street will be unable to tell if you actually entered the building or are merely hiding in the darkened doorway. This ploy draws anyone tailing you out into the open where you can deal with them more easily.

Use storefront windows, polished surfaces on vehicles, and side-view mirrors (when opening doors) to watch behind you or across the street. Eyeglasses and hand-held mirrors can also be used for this purpose.

Surveillance helicopters and other aircraft may be used to tail you from overhead. Equipped with night scopes and body-heat detectors (thermal imaging equipment), such aircraft are quite capable of tracking a suspect vehicle or a suspect on foot. Moving through buildings and under concrete structures (overpasses, parking garages) allows you to elude such flying tails, even those armed with heat detectors.

Using the Light Line
Unless in an area like Times Square, neon signs tend to be no higher than 20 feet up the side of buildings, and street lights do not normally go up to roof level. As a result, a light line is creat-

ed, usually halfway up buildings, leaving the upper floors, ledges, and roofs of multistoried buildings obscured in darkness.

Most humans see on line of sight, seldom bothering to look up or down. Ninja knew about this human failing and turned it to their advantage, sometimes hiding unnoticed for days in the rafters of a castle, waiting for the right time to drop down on an enemy. As a result, urban night fighters use ledges and roofs to move down a row of adjacent buildings, unseen and unsuspected. Therefore, you must always watch upper-level windows and roof edges, since one of the best ways to tail a person down a street is by moving from roof to roof along a block of adjacent buildings. Urban night fighters can also use alleys, basements, and even sewers to move unseen through an urban battlefield.

Urban Dress

Your urban dress should fit the time and place to provide you with camouflage without causing you to stick out in a crowd. Several versions of urban camouflage clothing are presently being marketed, all with various blends of gray and black hues arranged in the same basic design as jungle pattern camouflage. While this kind of dress might be useful in urban warfare, its use in cities is somewhat limited, since the cover it provides the night fighter is offset by its liability, should that same night fighter find it necessary to blend into a crowd.

The best urban night fighter dress is clothing that is functional and allows you to melt into the shadows, yet which does not make you stand out in a crowd for those times when you must mix into a crowd to survive. Your best bet is to wear a reversible, hooded jacket that fits the cloaking needs of the mission and is easily discarded in favor of casual clothing underneath. This will allow you to quickly lose yourself in a crowd. A pair of utilitarian coveralls, appropriately colored and easily discarded, might best fit the bill.

Obstacles

Obstacles are a natural part of any environment. In an urban setting, you run into obstacles that are not found in a rural set-

Illustration 12. Moving along a structure that is horizontally oriented.

Illustration 13. Moving along a structure that is vertically oriented.

KNIGHTS OF DARKNESS

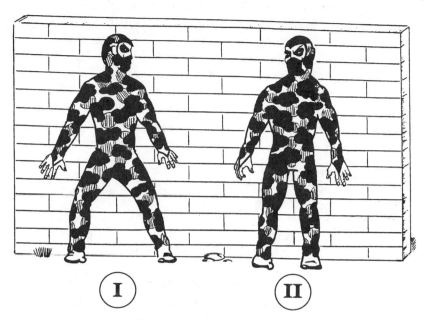

Illustration 14. The sliding step.

I: With your back to the wall, slide your lead foot forward a few inches just above the ground. Do not drag your foot along the ground, as this will make noise and leave evidence of your passing.

II: Placing the lead foot down carefully, slowly draw your rear foot forward in the same manner. Never place your feet too close together; maintain shoulder width between them. This distance gives you natural balance.

ting. In the city, lighting will be your greatest obstacle, but a good grasp of chiaroscuro will counter any obstacle of light that you run into. Moving along and over urban fences and walls of varying size and effectiveness also calls for specialized methods of movement not generally encountered in a rural setting. When moving along a wall, always stay as close as possible to the wall to cast a smaller shadow. Remember, the eye looks for contrast. The less you contrast with your background, the harder you will be to spot.

When moving along a structure that is predominantly horizontally oriented, use a crouching walk. Keep your hands as close to your body as possible to cast a smaller shadow and present less of a silhouette. An alternate method of holding your arms when moving along such a structure is to extend your arms to match the horizontal lines of the wall. Use the *half-moon* or the *kano* step (Illustrations 15 and 16 pp. 96 and 97).

When moving along a structure that is predominantly vertically oriented, maintain an upright walk that will better help you blend with the straight vertical lines of your background. When moving against such a structure, use the *sliding* step (see Illustration 14). Be sure to keep your knees flexed without dropping into an exaggerated deep stance. The more natural a posture and a method of moving, the more quickly and confidently you will be able to move through the darkness without being detected.

RURAL ("BUSH") MOVEMENT

Death doesn't mind hiding in the shadow of trees, no matter those trees be on a mountain trail in Colorado or along a jogger's path in Central Park.

—Dirk Skinner
Street Ninja

There are as many methods of moving through the countryside as there are types of countryside. All knights of darkness have their favorite techniques that work for them, but all night fighters share certain tactics and techniques.

You might be saying to yourself, "I live in the city. Why do I need to learn how to move in the countryside?" Consider Skinner's words: "Aren't there trees in your backyard? What about trees in your local park or playground? A simple rural consideration, such as the type of shadows trees give off, can spell the difference between your survival or your becoming just another urban mugging statistic."

Careful Movement
This is the key to moving through the countryside. You must

keep all of your movements slow in order to not attract attention. (Remember that the eye sees movement first.) When you do move, use every bit of cover (e.g., trees, large rocks, ridges, depressions, tall grass) available. Use body coloration and camouflage that helps you blend into your surroundings.

Should you "feel" danger, trust your sixth sense. Keep in mind that animals have a natural freeze reaction that helps save their lives when pitted against predators that hunt by sound and movement rather than by keen eyesight. For example, a bull isn't angered by the red cape, but is rather attracted to its movement. This is why the matador can stand still within inches of a raging bull that is attacking the fluttering cape. So, when you are crossing open fields, freezing in place can save your life and gain you precious seconds. Even a row of standing men may be mistaken for a row of trees, provided they are not moving.

Crouching in the open, you can be mistaken for a boulder or a bush. If you are being tracked by dogs, moving upstream can help, but this is no guarantee that the dog will be thrown off your scent, since your pheromones may have been blown to the stream's bank. Be careful when moving up streams that you do not turn over stones or leave footprints in the muddy bottom, which could leave a visual clue for human trackers. Also beware when exiting a stream not to leave wet tracks that will give you away. Conversely, you can purposely leave false exit tracks on shore to lead human trackers away.

Footfalls

How you step and move your feet can spell the difference between success and disaster, life and death. Novices always ask veterans how to walk through the bush. The vets answer in a deadly serious tone, "Very carefully!"

The most important thing is to find a method of moving your feet that is effective and natural. You must avoid any kind of complicated stepping method, techniques that call for you to cross your feet, and techniques that have you balancing precariously on the balls or heels of your feet. The trick is to choose a stepping technique that allows you to move as naturally as possible yet, at the same time, does not expose you to danger by slowing your ability to move when swift movement is all that stands between you and the Grim Reaper. The rule of thumb is to touch as little ground as possible without upsetting your balance. This means that you shouldn't crawl or roll on the ground unless it is absolutely unavoidable. Crawling and rolling bring more of your body into contact with the earth and increase the chances that you will disturb dust, rocks, twigs, and grass. Any good, sure-footed, flexible stance will allow you to move quietly through the bush while, at the same time, keeping your form low. This makes you less of a target and less likely to be spotted.

When crouching movement is called for, crouch low with knees bent only enough to allow you to get the full use of your "knee springs." Observe the "knee-toe" rule: Never let your knee extend beyond your toe. This position permits you to spring to either side should danger demand it. When you are closing with a human target, it is especially important to use a crouching stance that keeps you under the target's line of sight (see Illustration 26). Remember that humans tend to look around at a level that is constant—the level of their eyes. Unless something catches our attention—a noise or movement picked up by our peripheral vision that draws our attention either up or down—we continue to look along this eye-level plane.

Criminals have long known and put to use this line of sight concept. In his book *Thirty Years a Detective*, Allen Pinkerton reveals many of the tactics and techniques used by 19th century housebreakers and hotel thieves, which are still used by criminals today. In describing the methods of movement of hotel burglars, for example, Pinkerton details not only the crouching posture a criminal adopts when entering a room at night, but also explains the concept of line of sight. According to Pinkerton, the thief, when entering the room of a sleeping person, maintains a "stooping posture" as he moves to search the victim's clothing for valuables. If the victim stirs, the criminal quickly drops to his knees:

The reasons for adopting this stooping position or falling upon the knees is obvious, as every person in bed on being awakened suddenly, will naturally look up and not down. His movements are as rapid as lightning, and as noiseless as the Indian on the trail.[1]

Types of Stepping

Most night fighters use a type of half-moon technique that is similar to techniques taught in martial arts schools. For smooth, quiet movement that ensures solid footing and allows you to quickly shift from standing to crouching, this technique is best. Half-moon stepping will come as second nature to most night fighters who are familiar with Eastern martial arts, as this is the basic step taught by both karate and kung-fu instructors.

Wolfshirt warriors employed a stepping method known as *kano* stepping, which is similar to half-moon stepping. The similarities result from the fact that both are simply natural and effective ways of moving.

Kano stepping, also known as *ken* foot, was derived from the runic symbol that became the letter K in English. The rune *kano-ken* means *torch*, and this method of stepping, especially at night, was considered a light or torch to help one see and move better in darkness. Note the similarity between half-moon (a light in the night) and kano-ken (a torch).

Kano stepping differs from half-moon stepping in the way it shifts your weight. In kano stepping, the body does not stay in a direct line of forward movement, as it does in half-moon stepping. Although the kano practitioner moves forward, his body is constantly shifting from side to side, similar to the bob-and-weave of Western boxers. This type of forward movement, which employs a deceptive side-to-side movement of the body, made it difficult for someone to draw a bead on the wolfshirt. It also allowed the wolfshirt to attack an enemy (in the original wolfshirts' days, such enemies were archers and spearmen) head-on, which made his zigzagging body nearly impossible to target.

Illustration 15. Half-moon stepping.

I: Beginning in an even-weighted stance (50 percent of your weight on each foot) shift your weight to your lead foot, bringing your rear foot forward and in toward the lead foot and then out to a position mirroring your previous position. The feet have now exchanged positions.

II: Repeat this motion with the opposite side foot. Viewed from above, the movement of your feet resembles half of a circle or moon.

KNIGHTS OF DARKNESS

Illustration 16. Kano-stepping.

I: Beginning in an even-weighted stance, draw your rear foot in sharply at an angle toward your lead foot and then step out and forward to assume the same position you began from, albeit on the opposite side.

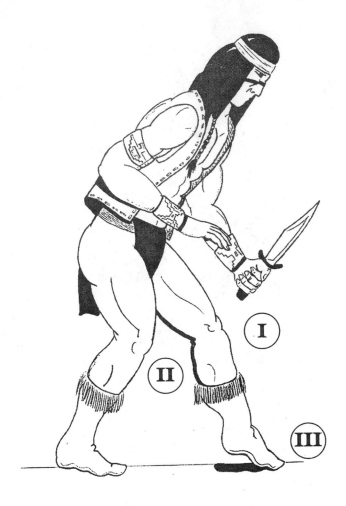

Illustration 17. The deerstalker step.

I: In a slightly crouching posture, slowly lift your lead leg's knee, allowing the lower calf and foot to dangle loosely (without tension).

II: Slowly swing the lower part of your leg forward while keeping the knee suspended in place. As the lower part of your leg swings out, toe forward, slowly lower it to the ground by lowering the knee.

III: The toe of your foot touches down first and carefully tests the ground to ensure that the rest of your foot does not lower itself onto rocks or twigs that could make noise. After the toe touches down, the rest of your foot follows slowly, beginning with the front of your foot and proceeding toward the rear of your foot until your entire foot is firmly settled. The process is then repeated with your opposite foot.

           KNIGHTS OF DARKNESS

Since Pinkerton saw fit to praise the stealth of American Indians, it is worth noting that, despite myths to the contrary, most Native American tribes had few reservations when it came to fighting at night. In stepping, Native American night fighters used a straightforward deerstalker step. The Indians (and later mountain men) who used this method wore either thin-soled moccasins or went barefooted so they could fully feel where they were setting their feet down, something not possible when wearing modern, thick-soled footwear.

SPECIALIZED METHODS OF MOVEMENT

These methods abound, as do tricks of movement.

Disguising Numbers

Many night fighters practice walking in each other's footprints. This helps to disguise the numbers in a team in case a foe comes across your tracks. The same practice is used by modern special forces to help cut down on the number of casualties caused by booby traps.

One ninja ploy involves a teammate carrying a second teammate on his back. This confuses the enemy as to how many are in the ninja team. Since the weight of a man can be determined by how deep a footprint he leaves, carrying a teammate on your back can fool experts as to the weight and size of the person making the tracks. Also, wearing women's shoes can confuse an enemy as to the gender of the person making the tracks.

Disguising Direction

The old trick of walking backward to disguise the direction of movement can still work on a limited basis. Ninja developed special shoes that left reversed footprints. They also used a special cross-step that disguised their direction of movement.

The Art of No Movement

When it comes to night movement, the most important kind of movement is no movement. Police stakeouts, ambushes, and

Illustration 18. The ninja cross-step.

I: Standing in an even-weighted stance, lower your center of balance while crossing your trailing leg behind your leading leg.

II: Having completed your cross-step, your crossed-behind foot points in the opposite direction you are moving, making it difficult for anyone coming across your tracks to discern your direction of movement.

III: Having completed the cross-behind movement, shift your weight to your rear leg and move your leading leg forward. Repeat.

hiding all require patience and the ability to remain motionless and unseen, often for hours or even days at a time. This is where your relaxation and meditation training comes in.

ENDNOTE

1. Allen Pinkerton, *Thirty Years a Detective*, (New York: G.W. Carlton & Co., 1884).

KNIGHTS OF DARKNESS

CHAPTER NINE

Reconnaissance

The same darkness that conceals movement also conceals obstacles and pitfalls.

—David Lovejoy
"Night Ordeal," *Gallery*

Before embarking on any night mission, you must first make a thorough reconnaissance of your AO. You must also examine the route you will take to and from your target.

The purpose of a reconnaissance is to gather all information possible about your AO to increase the probability of mission success. A proper reconnaissance will tell you when and how to move against your target. Conversely, a half-assed reconnaissance will result in failure and the imprisonment or death of you and your team. To conduct a thorough reconnaissance, simply apply those lessons that you were taught as a child about what to do before crossing the street: stop, look, and listen.

Proper reconnaissance answers two vital questions: What is the objective, and what obstacles are likely to interfere with our reaching that objective? When doing your recon, use all your senses. Look, listen, smell, taste the air, and get the feel of the AO.

Your reconnaissance should establish possible sights for your remain overnight (RON) bivouac. In an urban setting, your RON consists of establishing inconspicuous safe houses. In a rural operation, your RON consists of setting up a secure camp perimeter each time you stop.

A thorough reconnaissance is also vital for identifying potential obstacles. Clearly defining your objective and identifying all obstacles between you and that objective is the crux of the night fighter's craft.

OBJECTIVES

What is your goal? What is it you plan to accomplish?

Objectives fall into two categories: intelligence gathering (e.g., the gathering of information, acquiring of documents, surveillance of enemy personnel, capturing of knowledgeable individuals) and interdiction (e.g., interfering with your enemy's plans through guerrilla activities, sabotaging). Whatever your objective, it must be clearly defined before starting on a mission. This cannot be emphasized enough. More than one mission and more than one war have been lost simply because of a lack of a clearly defined objective. (Failure to clearly define your objective sows the seeds for debacle, disaster, and conflicts like Vietnam.)

Each member of the team must know the part he is to play and, in case of emergency, be able to carry on for a dead or wounded teammate.

Intelligence Gathering

If the objective is intelligence gathering, contact (i.e., confrontation) with the enemy must be avoided. Losing sight of the objective of gathering information by needlessly attacking sentries jeopardizes the mission and may compromise or negate the value of any intelligence that is gathered. Information is most useful when an enemy doesn't know you have access to that information.

Interdiction

Likewise, if your objective is interdiction, all of your efforts must be concentrated on upsetting the enemy's plans or inflicting as many casualties upon him as possible. Should intelligence fall into your lap during an interdiction operation, it can be pocketed for examination at a later date.

In a major operation, two teams, one with the objective of intelligence gathering and the other concerned with interdiction, will plan strategy in sync.

OBSTACLES

Obstacles to your intelligence gathering or interdiction fall into one of three categories: terrain to be crossed, security perimeters to be circumvented, and structures to be penetrated. Common to all of these is terrain, of which there are two basic types: urban and rural. All three types of obstacles offer different challenges to the night fighter.

Urban Terrain Reconnaissance

In performing a reconnaissance of an urban AO, you must plan such things as your primary routes into and out of the area, as well as alternate routes. It is important that you establish these routes beforehand. When possible, reconnoiter the proposed routes before the mission, during the day perhaps, in anticipation of a night mission. When planning your urban route, take into consideration the vehicles that will be used. Vehicles should be chosen to blend in with others in the AO. During the operation, observe traffic laws while driving. (Any good reconnaissance should take the local authorities into consideration—it would not do for a group of night fighters to be pulled over because of a faulty taillight!)

Use the sound of vehicles being started to hide your movement. How often is a gunshot mistaken for the backfire of an auto?

Places for parking or otherwise hiding a vehicle should be established during your recon, as should secluded spots for abandoning stolen vehicles or switching getaway cars. Do your vehicles need to be disguised (e.g., false license plates, identification stickers, magnetic business signs affixed to the doors, police lights on the roof)? The silhouette of a vehicle can be easily disguised. At night, any car with a rack of red or blue lights on the roof can be mistaken for a police car. Pulling up behind a car on a darkened road, your headlights on bright, all anyone in the stopped car can see is your flashing red light. When walking up to such a stopped vehicle, keep the bright beam of your flashlight shining directly into the car, which will keep you in silhouette. The addition of obvious police headgear, a protruding sidearm, and a nightstick will further disguise your silhouette and complete the masquerade.

Are the roads your vehicle will travel toll roads? Are they prone to heavy traffic that can either slow your getaway or hide you? Will you be traveling on one-way streets? How will you dispose of the getaway vehicle afterward? All of these questions must be answered.

If you are not using a vehicle, you need to gauge your walking route to take advantage of shadows and blind spots for concealment. Careful examination and calculation should be made to determine where moon shadows will be cast by structures at various times. Recon the area to decide the best clothing to wear (e.g., military camouflage, ninja black, casual). Note any special type of dress the inhabitants of the area wear in case you need to mingle with them at some point or disguise your silhouette to fit the area.

When applicable, you must also consider where you will bivouac if your mission takes more than one day or night. In an urban AO, a safe house needs to be set up that will act as both a rendezvous and staging area and a communications post. (In a rural setting, a secure camp perimeter must be set up each time you stop.) Safehouses and other urban bivouacs should be set up ahead of time. Alternative routes must be established to such a rendezvous.

During your initial reconnaissance, take time to "soften up" the area. Also known as ground laying, the softening up of an area can involve anything from planting weapons and other pieces of support equipment to disabling lighting. And any good recon, whether urban or rural, must take weather into consideration. In an urban setting, weather will influence the number of people on the street. Ice- or snow-covered roads and streets can make for difficult driving and leave telltale tracks. Always have a "plan B."

Rural Terrain Reconnaissance

Recon in a rural environment bears many similarities to recon in an urban environment, and it is influenced by such factors as distance, the availability of cover for movement and bivouac, and the effects of weather.

Weather conditions (current and what is forecast for the time of the operation) are a crucial factor. A mission will often have to be scrubbed because weather conditions refuse to cooperate. On the other hand, cloudy weather at night can obscure the moon, providing extra cover for moving through the countryside. However, no moon at all can make movement in the countryside more difficult. Inclement weather can throw off pursuers, but it can also bog you down.

When crossing rural terrain, be careful to pick a route that takes full advantage of shadows and protective cover. Beware of startling animals and beware of animals that might startle you. (Modern man is out of place in the bush. Animals will either scream about his presence or give him away by suddenly going silent. A thorough reconnaissance of the intended AO should include native animals. Team members can learn whistles, croaks, and other sounds of native species in order to communicate without giving away the team's presence and to recognize an enemy using the same trick.)

Make sure all equipment is taped down or otherwise secured to prevent unwanted noise. Avoid coughing, sneezing, and flatulence, and don't move along ridges or against lighter colored backdrops (e.g., hillsides) that silhouette you. Always move at irregular intervals. Never give your enemies a routine of movement that they can turn against you. When moving, stop, look, and listen for half the time you move. Move 20 minutes and then listen silently for 10. Establish point and rear guards when moving as a team.

Rather than rely on speech, establish hand signals to supplement the aforementioned natural sounds to communicate with team members. American sign language is a valuable skill for all night fighters.

When moving through wooded areas, replace branches carefully. Never let branches snap back and make noise or injure the eye of a team member. Whenever possible, use the natural sounds in your AO to mask your movement. Use the sound of wind or thunder to disguise your activities. For example, thunder can muffle the sound of firearms and small explosions. During combat, use the sound of artillery or gunfire to disguise movement.

Stay off well-beaten paths, but do so without needlessly stumbling through the bush. Cross open spaces one at a time. If you believe you have been spotted crossing a wide, open space, freeze in place rather than try to dash to cover. A squatting man silhouetted in an open field, even in broad daylight, will often be mistaken for a boulder, bush, or tree stump. Remember, eyes see movement first.

Scatter dead leaves and dried twigs behind your path to help prevent an enemy from sneaking up on you. (In buildings, ninja used rice for this same purpose.) Whether you have to cross varied ("broken") terrain or uniform terrain is also a vital consideration in your recon.

Crossing grasslands is different from crossing a swamp or a desert. Each type of terrain offers special assets and liabilities for night fighters, which a thorough reconnaissance must uncover. If there are bodies of water, you will want to note these in your recon. Some bodies of water cannot be avoided, in which case plans will have to be made to ford or otherwise cross them. Moving water can be used to facilitate movement, too. Standing waters are best bypassed because of their openness and reflective nature.

Moving through a heavily wooded AO, whether a forest or a jungle, calls for special preparation. Ultimately, there is no such thing as moving silently through the jungle at night, and many accomplished American woodsmen, who were used to the thick, dark woods of North America, found themselves out of their element when confronted with the pitch darkness of Southeast Asian jungles, where the overhanging canopy at times completely obscured all light:

> Jungle nights are so dark that a person will experience vertigo, for there is no reference for the human equilibrium system. The symptoms of vertigo are often accompanied by claustrophobia, compounding an already unstable situation. I have experienced the symptoms of vertigo and claustrophobia, both separately and in combination.
>
> —David Lovejoy
> "Night Ordeal," *Gallery*

One U.S. Marine Vietnam veteran, author John F. Johnson, recalls learning to move in pitch darkness through the dense bush by feeling the direction leaves were turned and checking to see which side of a tree was moist; by learning on which side of a tree certain moss grew, his touch compensated for his lack of sight. Other accomplished night fighters express similar respect for the challenges of night movement through the bush.

It is good practice to establish your RON just before dark and then move it shortly after dark. This prevents an enemy who may have spotted your group setting up camp from attacking in the night. Setting up a false camp (e.g., empty sleeping bags, a campfire) is a classic counterambush tactic. Pickets should be immediately established and should include a roving sentry to double-check on posted sentries.

Make sure any bivouac takes into consideration where shadows are at the time the RON is set up and where those shadows will be in a few hours. It would not do to set up a RON totally obscured in the shadows of evening only to have your position completely illuminated by the moon in the early hours of the morning.

Watch your waste. Do not leave any litter to mark your passing and cover all latrines. Avoid unnecessarily disturbing an area, such as breaking branches or pressing down grass. Be careful when using streams, because debris or even soap bubbles can float downstream and alert an enemy to your presence. The most important thing when you are moving through the bush is to concentrate on what you are doing, not wishing you were somewhere else.

Planning Your Route

Having determined the special requirements of the terrain, you must then factor into your recon the best routes through and around that terrain. This planning also involves how you will penetrate closely guarded security perimeters and structures. Professional cat burglars often penetrate a well-secured building over several forays, with each successive raid taking them a little farther into the security perimeter and closer to their final objective. These preliminary reconnaissance forays help them feel out the security procedures, make key impressions for locks, and plant equipment that will be needed for the actual job. At the simplest, this ploy involves visiting a targeted building during the day and unlatching a window, through which you will enter later that night.

In planning your route, take advantage of aerial photos, maps, fly-overs (e.g., hang gliders, parachutes), and eyewitness accounts of team members who may have operated in the AO before or on similar terrain. Designate rendezvous points in the event that team members become separated. Remember that a thorough reconnaissance of the AO should include mapping paths of least resistance, shortcuts, and alternative routes of ingress and egress.

Ironically, the most practical advice for night fighters moving through any area comes from a philosophical Zen adage, which advises us how to move through life in general: *seek passage without traces.*

CHAPTER TEN

Perimeters and Structures

As a night fighter, you might find yourself faced with various perimeter and structure obstacles, regardless of the AO. When avoiding such perimeter obstacles is not an option, you must find ways to go over, under, or through those obstacles.

PERIMETERS: THE THREE WAYS OF GOING

Going Over

Crossing over a perimeter can be anything from climbing over a fence or wall to using hang gliders or HALO (high altitude, low opening) parachutes. On buildings that are close enough, you can simply jump from the higher level building to the lower roof, thus completely bypassing a perimeter fence or checkpoint. In an urban setting, buildings will be close enough to allow shimming up adjacent walls by bracing your back against the wall of one building and your feet against the opposing wall. Heavy power lines can also be used when they are imprudently draped over a wall or fence. However, proper insulation must be worn or, if possible, the power supply turned off.

To climb over other walls and structures, grappling hooks

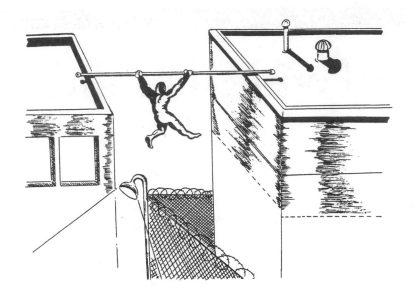

Illustration 19. Penetrating a perimeter above the light level.

You will recall that the advantage of using adjacent roofs and ledges to cross over a perimeter is that the tops of most buildings are shrouded in darkness, since they are above the light line.

can be thrown or shot across gaps, and traverse lines can be set up. In the absence of a grappling hook, you can use pieces of pipe, discarded pieces of stout wood, or a ladder to cross between buildings.

Modern fences offer certain challenges not found in medieval castle walls. Fences can be equipped with motion sensors designed to raise alarm at the slightest touch. Still other fences are profusely topped with barbed wire. The type of fence you are facing will determine how you attack it.

Modern night fighters are more likely to have to navigate a fence topped with barbed wire than a castle's walls. However, the same principles are applied to both. Fences come in all shapes and sizes. Chest-level fences can be vaulted, but taller fences (10 to 15 feet) require more preparation. Some tall fences have chicken

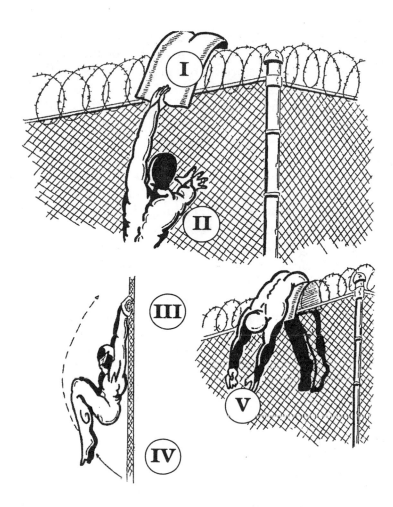

Illustration 20. Negotiating a fence.

I: Either through your own efforts or by standing on the shoulders of a teammate, swing a cloak, jacket, or piece of carpet over the barbed wire.

II: Beware of leaning against fences that are equipped with motion sensors or electrified.

III: Hook your hands into the fence above your head, palms toward the fence, elbows pointed forward.

IV: Drawing your legs up to your chest, in one smooth movement arc your body up, back, and over your cloak. This movement is similar to doing a chin-up.

V: As your body slides over the top, release your grip on the fence and grab your cloak, pulling it after you.

wire near the top. This finely meshed screen requires the use of ninja *nekode* ("cat's claws") or other tools for hooking and drawing yourself up to the point to where a grip can be had on the top of the fence. For fences without motion sensors, a simple hooked cane can be used to draw yourself up or pull down barbed wire. Other types of fences have angled tops designed to collapse inward or outward (depending on whether the fence is designed to keep invaders out or prisoners in).

Never leave anything behind that will betray your breach to a passing sentry. As you drop to the other side of the fence, be careful about your footing. Irregular-shaped rocks and soft sand are spread at the foot of security fences to make it more difficult for intruders to get their footing and cause them to make more noise; the soft sand causes them to leave tracks.

Going Under

Going under a perimeter can be as simple as wiggling under a fence or as complicated as spending months digging a tunnel through dirt and rock. (Tunneling has been used down through the ages to undermine castle walls and escape from POW camps.)

When dealing with flexible (moving) perimeters, one of the best methods is for sappers to anticipate enemy troop movement and dig tunnels and covered foxholes ("bolt holes") where it is anticipated the enemy column will bivouac. At least once during the Vietnam War, an American army base was inadvertently built directly over a Vietcong tunnel complex. Whether this occurred because of coincidence or the Vietcong's anticipating the Americans' intent to build a camp on the site and then simply extending their vast system of tunnels to the site, the result was the same: enemy sappers appeared and disappeared at will within the perimeter of the base.

In modern times, the North Koreans have made a habit of digging tunnels under the heavily guarded 38th Parallel. Criminals have likewise used tunnels, those provided by public works and those they industriously dug themselves, to break into banks, including the Bank of England.

Streams and rivers running under fixed-perimeter fences can

be navigated to gain entrance. Such waterways can also be used to silently approach enemy RON positions that are set up near rivers.

All installations require water, electricity, sewerage, and other utilities. Maintenance tunnels and access conduits running under perimeter walls and fences can be slithered through. To demonstrate this fact, from medieval Japan comes the true story of "The Littlest Ninja," a diminutive night warrior who succeeded where his taller compatriots failed. The story goes that a certain shogun had repeatedly escaped assassination, despite the best efforts of the ninja clan he had offended to dislodge him from his seemingly impenetrable castle. Finally, this little ninja, a warrior who, because of his size was seldom chosen for clan missions, volunteered for the assignment.

First he found the master builder responsible for the construction of the shogun's fortress, and, through one form of persuasion or another, the dwarfish ninja secured the architect's plans for the castle. Within a fortnight, the ninja was to be found poised under the waste hole in the shogun's toilet, having slithered through several claustrophobic yards of shitty drain. No sooner did the shogun hear the call of nature and lower his ass over the toilet than he felt the ninja's long dirk penetrate up into his body.

Mission accomplished.

Going Through

A perimeter can be a daunting prospect when we first find ourselves confronted with such things as walls, barbed wire fences, bright lights, sentries, dogs, and a plethora of electronic sensors. But if a man built the lock, then another man can build the key.

True, fixed-perimeter obstacles can be climbed over or dug under, but they can also be passed directly through. Walls can be broken through and fences can be cut. Fences in disrepair can be pried up and away from supporting posts.

If cutting or digging through fences and walls isn't possible, the remaining option is to enter the perimeter through the front gate. Constructing a workable disguise (complete with proper uniform, arms, and, if need be, identification) that allows you to pass unchallenged through perimeter gates and checkpoints is always preferable.

Knights of darkness can enter a perimeter or building during the day, hide, and then emerge after dark. This is an excellent ploy when working as a team, since a single inside man can clear the way (e.g., opening windows or doors) for dozens of infiltrators.

Regrettably, there will be times when force must be used, when finesse is not a viable option. In such cases it will be necessary to remove any perimeter sentries standing between you and your objective. Unavoidably, this increases the possibility that your penetration will be discovered. To avoid detection, have a team member with a disguised silhouette replace the sentry. If working alone, prop the sentry up to make it appear as if he is still alive. In times of war, the rule is to never leave a living (even if unconscious and bound) sentry behind your ingress. Having removed the sentry, it is possible for a lone night fighter to don the sentry's uniform to disguise his silhouette to further penetrate a perimeter.

Another tried and true method for breaching a perimeter is to hitch a ride on a horse—a Trojan horse. Everyone is familiar with the story of the Trojan horse, how, after 10 years of stalemated warfare, Greeks besieging Troy tricked the Trojans into taking a large wooden horse filled with Greek warriors inside the walls of the city, with predictable results. Likewise, using the moving shadows of a vehicle by hitching a ride on the side, top, or undercarriage of a vehicle that is entering a target area is a sound modern application of the Trojan horse ploy.

The enemy within a perimeter requires that food and other goods be brought in on a regular schedule. Traffic in and out of an installation's gate provides ample opportunity for infiltrators to operate, especially at night. One variation of the Trojan horse involves hiding inside a pallet of boxes that, individually, are too small to hold a person. This kind of illusion is often used by professional magicians and is believed to have been invented by the moshuh nanren.

On a pallet of stacked boxes, the night fighter prepares a series of fake boxes at the bottom of the stack. Each of these

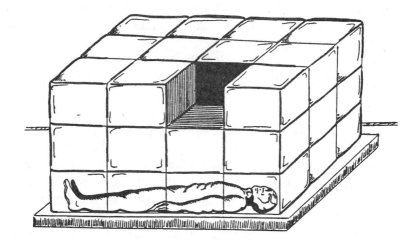

Illustration 21. The Trojan horse variation.

boxes alone is too small to hide a human being, but when arranged together in a series, they create a space that will hold a man. Boxes holding real materials are placed on top of the pallet stack so as to pass cursory inspection.

The boxes must be placed on a portable platform pallet so that they will be moved en masse by a forklift or a pallet-jack trolley, rather than one at a time. This ploy can be used to place an operative within a perimeter to allow him to emerge at the crucial moment.

There are endless variations of this trick, a ploy that plays on the perception of the human eye.

THE DETAILS OF PERIMETER PENETRATION

Perimeter penetration demands a careful study of both types of perimeters, as well as the security precautions and zones likely to be encountered when penetrating a perimeter.

All installations with security perimeters are divided into an outer zone (leading from the tree line up to the perimeter wall or fence), middle zone (the wall or fence itself), and inner zone, the latter of which consists of the exteriors and interiors of structures.

Types of Perimeters
Perimeters come in two types: flexible and fixed.

Flexible
Flexible perimeters are temporary perimeters, such as those set up overnight by a moving troop. These perimeters are guarded by pickets of stationary or patrolling (sometimes both) sentries and perhaps dogs. In times of war, flexible perimeters often consist of rings of foxholes, trenches, and barbed-wire barriers. They will usually be oval in shape, unless the camp has foolishly been set up with its rear to a body of water or unless it is laid flush against the back wall of a hillside or canyon. (A flexible perimeter set with its "six" [rear] to a canyon wall or a body of water is an infiltrator's dream, allowing infiltrators to use scuba gear [closed-circuit if possible] to swim up to or repel down into the camp.) Flexible RON perimeters will be strung with temporary alert devices, such as trip wires connected to everything from noise-making tin cans and flares to claymore antipersonnel mines.

In many ways, flexible perimeters are harder to infiltrate than are fixed (fenced or walled) perimeters. Fixed-perimeter guards soon grow bored, bogged down in the monotony of their nightly routine. Flexible-perimeter sentries, on the other hand, tend to be more alert, since they are guarding their lives and the lives of their bivouacked comrades.

Infiltrating flexible perimeters requires the utmost in patience and stealth, because it is a matter of sneaking past or silently removing human and sometimes canine sentries, as opposed to circumventing electronics on a fixed perimeter.

Fixed
Fixed perimeters are permanently set in place. They run the gamut from secured office buildings without fences to penal and military institutions with fences, patrols, and a host of electronic security devices.

Although generally having better overall security measures in the form of fixed obstacles (fences, electronics), fixed perimeters are generally less secure in terms of personnel. There is an inverse relationship between the amount of high-tech security equipment and the alertness of perimeter guards. In other words, the more you depend on your high-tech equipment, the less you will trust and depend on your own senses.

A thorough reconnaissance of any perimeter you intend to infiltrate should include both a ground-eye view and, when possible, a bird's-eye view. This recon must assess the quantity and quality of both the personnel—via analysis of their movement patterns—and the types of security devices protecting (e.g., lights) the perimeter. The ground-eye view can be through the study of maps and floor plans, eyewitness accounts, and, where applicable, visits to the perimeter during open hours (when the building or base is open to the general population). In the latter instance, however, you must always disguise yourself when reconnoitering a perimeter, since after a breach of security (e.g., night infiltration, robbery), perimeter security personnel review video surveillance taken during the days prior to the break-in, looking for suspicious individuals caught on film casing the joint.

Guards and dogs working a perimeter will unintentionally reveal where security devices are located by where and how they walk (or don't walk) along certain prescribed paths, while avoiding others.

For your bird's-eye view, use aerial photos or do a fly-over. A lone hang glider or even a hot-air balloon passing over a perimeter, although cause for curiosity, need not necessarily be cause for alarm. When this is not possible, get a good look at a fixed perimeter from nearby hills, buildings, or radio and water towers.

Types of Security

Security perimeters fall into one of two categories: active and passive.

Active

This consists of mobile security personnel, such as sentries, dogs, and robots.

No matter how sophisticated the security device, all have one intrinsic flaw: they depend on human beings to respond to their electronic cries for help. In the final analysis, humans remain your greatest obstacle—and your best ally.

Some guards are highly trained professionals. These are the high-caliber security personnel found at high-security military installations or those guarding Medellin drug cartel interests. From this high standard of professionalism down, perimeter guards fall somewhere on a descending scale of efficiency ranging from the chronically bored to the criminally corrupt. The rise in the overall education level of private security personnel (in 1971, the average education level was ninth grade) is not a cause for concern by infiltrators. Why? Because the average education level of private security personnel has risen in the past few years due to the fact that many unemployed, albeit well-educated peo-

Illustration 22. Security precaution zones.

I: Tree line: the farthest extent to which the security perimeter is designed to control.

II: Exterior fields of fire: the open ground cleared of vegetation and obstacles (e.g., large rocks) between the tree line and the beginning of passive and active security measures.

III: Outer zone: the area where actual security devices and personnel are encountered. It can consist of flexible or fixed precautions, sometimes both.

IV: Breach barriers: the walls or fence(s) surrounding an installation that are often augmented with alarms and sensors and fortified with sentry boxes and guard towers.

V: Inner zone: this zone contains sensors and patrols similar to those found in the outer zone. Additional considerations include free-roaming dogs and an increase in "electronic trip wires" (sensors using infrared beams, pressure mats, photoelectric motion detectors, seismic or ultrasonic sound detectors, and microwave beams).

VI: Interior fields of fire: the cleared ground between breach barriers and the interior structures (e.g., buildings, aircraft) being guarded.

VII: Structure interior: the exterior walls of targeted buildings and structures within the perimeter itself. They are guarded by electronic, human, or canine security, or a combination of these. They may present reinforced entrances, locks, and false entrances designed to confuse infiltrators.

VIII: Structure interior: the interior of buildings may be guarded by electronic trip wires or by active security. In addition, stores of information or valuables may be secured in vaults or reinforced cabinets that are also equipped with proximity or motion sensors.

KNIGHTS OF DARKNESS

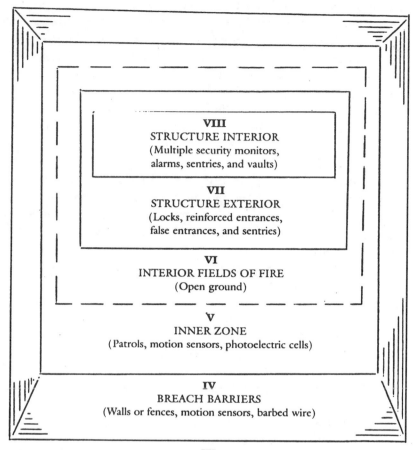

VIII
STRUCTURE INTERIOR
(Multiple security monitors,
alarms, sentries, and vaults)

VII
STRUCTURE EXTERIOR
(Locks, reinforced entrances,
false entrances, and sentries)

VI
INTERIOR FIELDS OF FIRE
(Open ground)

V
INNER ZONE
(Patrols, motion sensors, photoelectric cells)

IV
BREACH BARRIERS
(Walls or fences, motion sensors, barbed wire)

III
OUTER ZONE
(Patrols, motion sensors, pitfalls, and land mines)

II
EXTERIOR FIELDS OF FIRE
(Open ground cleared of vegetation and obstacles)

I
TREE LINE
(Outward extent of perimeter security)

ple, have been forced to take lower-paying jobs as rent-a-cops just to make ends meet. However, rather than raising the overall efficiency of private security personnel, this influx of better-educated workers has had the *opposite* effect on the efficiency, because many of these unemployed college graduates resent the fact they have had to take low paying, less-than-prestigious jobs.

It is important for night fighters to understand the psychology behind this resentment, as well as other conditions (e.g., thoughts of promotion, personal conflicts between fellow workers) that might be used to motivate (read: manipulate) security personnel. For instance, security personnel assigned to guard a perimeter, especially a fixed perimeter, are vulnerable to bribery, boredom, sloth, confusion, disinterest, threats made against self and loved ones, and death, any of which you can employ.

A thorough reconnaissance of a perimeter should include observations and, when prudent, physical (health) and psychological profiles of security personnel. Do the security guards appear alert? What physical condition are they in? Could they be easily overcome? Will they give chase? Do they smoke or drink? Could they be (or are they already) drugged? Could their uniforms and ID be stolen? Do they have families who can be manipulated? Are they honest, or do they pocket a little here, a little there each night? Do they sit on their asses, or do they roam around more out of boredom than out of any actual concern for security?

When doing your reconnaissance of security personnel, mark both their attitudes and their routine. Are they required to make timed rounds (to punch a watchman's clock)? Are they in fixed towers or sentry boxes? Are these boxes vulnerable to outside attack? If not, what is the possibility of using a Trojan horse ploy? How are the guards armed? Do they have firearms or simply alarms? Are the alarms on their belts or at a fixed location on an out-of-reach wall? Are their guns real? (Some private security services only provide their guards with fake ["counterfeit"] firearms to cut down on the costs of bonding armed security guards.) Are guards hidden behind one-way mirrors or bullet-resistant glass? Do they have night scopes or other sophisticated equipment? They may or may not have been trained to use this equipment properly. For example, are the guards equipped with dogs? Do they know how to handle the dogs correctly, or are the dogs left in an enclosed area to bark? You can tell a lot about an installation by the fact that there are dogs patrolling a perimeter. Dogs controlled by trained handlers (on a leash) indicate a perimeter that is serious about security. On the other hand, a perimeter where dogs are permitted to run free indicates a decidedly cheaper concern for security. (Dogs have been used since ancient times to protect property, ever since the first domesticated wolf-dogs were given the trusted spot guarding the mouth of our primitive ancestors' caves. Dogs are still used today, often because they are cheaper than human guards.)

Guards who are left to run helter-skelter inside a perimeter indicate the absence of sophisticated motion detectors and beam trips, sensors that might accidentally be set off by wandering guards and animals. Beware of trip beams placed at waist height—high enough not to be tripped by dogs, yet low enough to be tripped by humans. This is also a pretty good indication that the fence itself (at least the inside fence, if there is one) is not wired for electric shock nor equipped with pressure-sensitive or motion-sensitive alarms, since a dog jumping against the fence would trigger such alarms.

Dogs, while initially intimidating, are easily dealt with. They are vulnerable to most of the same measures used against human sentries: distraction, bludgeoning, poisoning, and silent weapons. In addition, dogs can be disabled by a host of ultrasonic devices, antidog (repellent) chemicals, and bitch scent. A simple ploy has one team member drawing the dogs to one side of the compound, and perhaps the dogs' master as well, while his fellow night fighters slip over the wire at the opposite end of the compound.

Lately, electronic "dogs"—robots—are all the rage. The robots presently in use consist of mounted cameras that follow a preprogrammed route through a facility or are manually driven from a remote console by security personnel with a joy stick. Such robots are generally found inside buildings, move slowly, and ultimately depend on the efficiency of their human controllers to respond to their alarm.

Along with electronic dogs we are seeing an increase in the use of computer monitors. In general, these computers take the place of the lone security guard who is in charge of watching multiple closed-circuit television (CCTV) monitors for anything out of the ordinary. To use a computer monitor, fixed-perimeter scenes are digitized, and the scene angles, contours, and shadows are fed into the computer as data. Should the original scene change (e.g., an intruder enter the picture), the computer alerts its human master. As challenging as this might sound to novice knights of darkness, the bottom line remains that even if an "infallible" computer spots you, eventually it will be an all too fallible human being who comes to investigate or attempt a capture.

Passive

Passive security consists of fixed barriers, such as fences, walls, locks, sensors, monitors, alarm trips, and traps. But don't let the name fool you: passive security can be just as dangerous as active security. However, any passive security, no matter how sophisticated, can be bypassed or otherwise overcome.

Passive security precautions consist of two types: barrier and alarm.

Barriers consist of walls and fences, lighting, locks, and interior secured areas, such as vaults or metal cabinets. Each one of these barriers can be further augmented with various alarms that are triggered when the barrier is approached, touched, or otherwise interfered with.

Illustration 23. Negotiating a wall.

I: Having gained the top of the wall, keep your body tight against the structure, molding the contours of your body to it.

II: Do not drop equipment or throw it over the wall. Instead, lower it quietly with a cord. Even when working in a team, equipment should never be tossed over an obstacle. Even a ninja can miss a catch.

III: Slowly slide your body over the edge of the wall, retaining your grip with one hand and foot while helping control your slide down the wall by keeping your leading hand and foot pressed against it. Avoid scrapping sounds and leaving skid marks from your shoes, which could alert a passing sentry to your passage.

IV: Release your remaining foothold on the wall, momentarily dangle (suspended by your hand), release your grip, and drop the rest of the way to the ground.

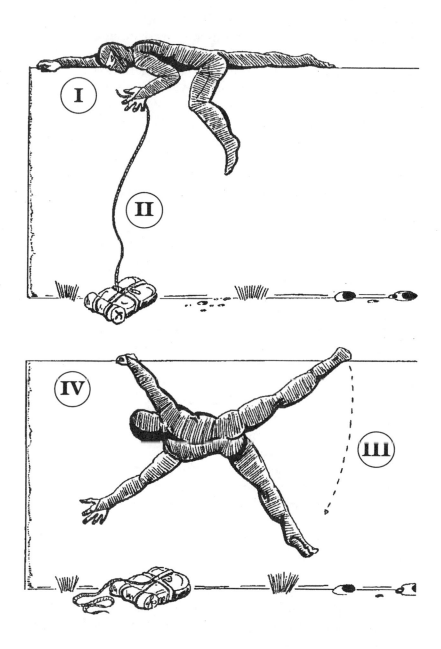

The walls that an infiltrator will have to contend with will be those surrounding the perimeter itself and those to a structure that he must first penetrate and then move through.

Some of these walls must be climbed. When you are climbing over a wall, or any obstacle for that matter, the key is to spend as little time as possible silhouetted on top of the wall.

Landing is as important as climbing. When dropping to the ground, always land in a crouched position with your knees bent to absorb any shock. When dropping from a greater height, as you feel your feet touch the ground, fold your legs into a controlled collapse as you fall forward into a forward shoulder roll (your weight passing diagonally across your back as you roll). Come up in a combat-ready squat. Be careful not to land on any of your equipment or teammates who have preceded you. Be sure there is no debris (e.g., rocks) that you might inadvertently land on and cause noise or twist your ankle.

When faced with a wall that is adorned on top with imbedded glass shards or barbed wire, adapt the method in Illustration 19 (p. 110).

Other walls may have to be circumvented by digging under or breaking through them. Exterior walls, such as those surrounding a perimeter, although solid, may be topped with barbed wire or sensors. Some walls will be impregnated with seismic devices designed to "pick up" anyone trying to bust through them. These types of walls will mainly be found inside buildings, primarily in ceilings, walls, and vaults housing valuables.

Fences around a perimeter are often impregnated with extra security devices, setting off alarms when touched or moved. Also, lighting is a major barrier because shadow and silhouette are their stock and trade.

Electronic sensors rigged to alarms are many and varied. They can be affixed to walls, fences, and doors. They can be buried underground or consist of CCTV or listening devices.

Within a structure, part of the security perimeter will consist of vaults and steel cabinets that are designed to resist break-in. Many of these will be equipped with proximity and contact

MOTION TYPE	CONCEPT	ADVANTAGES	DISADVANTAGES
Fence-mounted sensor	Detection depends on movement of fence	Early detection on interior fence, follows terrain easily	Frequent false alarms (weather, animals), vulnerable to animals
Underground seismic cables	Sensor registers ground vibrations	Good for use in uneven terrain	False alarms due to heavy vehicles, etc.
Balanced level capacitance	Unbalanced level (mercury) sets off alarm	Few false alarms, good for use on rooftops, any terrain	Vulnerable to animal ploys and inclement weather
Taut wire	Change in wire tension sounds alarm	Few false alarms, good for any terrain, good for interior fence	False alarms from animals, ice, and snow; temperature changes require adjustment
ENERGY TYPE			
Microwave beam	Breaking of beam (volumetric)	Few false alarms, ease of maintenance	False alarms from animals, inclement weather
Infrared beam	Breaking of beam	Good for short distances (doorways, halls, etc.)	False alarms (animals, fog, dust, voltage surge)
Electric field	Penetration of volumetric field	Good on uneven terrain, free-standing or fence mounted	Vulnerable to animals, electrical interruption
Video motion detector	Depends on change in scene-image being monitored	Augments other systems	Vulnerable to electrical interruption, electrical storms, lighting change

Illustration 24. Security zone precautions assessment.

alarms. The two main types of such sensors are motion-triggered alarms and energy fields.

Motion-triggered alarms appear on fences as motion detectors that are triggered when someone jiggles the fence. One such motion sensor is the balanced capacitance sensor. Simply put, this sensor is a balanced mercury vial that, when tipped to one side or the other, activates the alarm. Another perimeter fence alarm uses taut wire. In this instance, the wire of a fence is drawn tight so that any change in the tension of the wires (e.g., cutting) triggers an alarm.

Energy fields, on the other hand, consist of sensors between which beams (microwave, infrared, photoelectric) pass. Any thing or anyone interdicting this beam will set off an alarm.

Other alarm sensors consist of video cameras activated by sound or motion. Seismic sensors may be placed in the ground near fences or in walls, floors, or ceilings within buildings.

Although initially daunting, sensors still have weaknesses that can be exploited. No matter how good a sensor is at picking up an intruder, in the end it still comes down to a human being sent to investigate the machine's cry for help. If you can make the human monitors doubt the efficiency and dependability of their machine counterparts, you will have set the stage for your perimeter penetration. Statistics indicate that 95 percent of all activated perimeter alarms are *false* alarms. In the case of a complicated security perimeter using several types of security sensors, the false alarm rate is expected to be at least one false alarm per day per security zone. This might not sound like much but, when you consider that a fair-sized installation's perimeter can have four or five security zones, that installation can expect four or five false alarms a day, every day.

How are security personnel to know when an alarm is true or false? The answer is that they don't! If the security personnel are highly trained, they may react to each alarm with enthusiasm, assuming that each is a real alarm. On the other hand, if the security guard is being paid minimum wage, he or she will undoubtedly react to every alarm with the assumption that it is yet another false alarm—another machine crying "wolf."

The more complicated an electronic security system, the more wires it requires and the more energy it uses. Wires can be cut or rerouted. Energy sources can be destroyed. A general blackout of power in an area will be an inconvenience to a security perimeter, but will probably not be seen as a direct attack on the facility itself.

Once triggered, most alarms must be reset. This often necessitates calling specialized workmen out of their sleep. Most alarm systems, especially those monitored by the police or a private security firm, are rigged through existing telephone lines, saving security firms the cost of having to lay their own lines.

Every machine, no matter how sophisticated, has a weakness. Even computers, for all their uses, are only as smart as the people who program and operate them. Likewise, you must remember that security devices are only as good as the people who install, maintain, and answer them. For example, if you force a door to which a trip alarm is attached, there is a pretty good chance that the security guards answering the alarm will know someone has forced the door. On the other hand, should the same guards respond to a photoelectric beam being broken and find a cat walking nearby, they will assume that "the stupid machine" has made yet another mistake.

Photoelectric cells, between which a beam of light bounces, can be circumvented by going over or crawling under them. Ultrasonics depend on an intruder making enough noise for them to register and are often turned off around parking areas and along stretches of road near perimeter fences traveled by heavy vehicles. The same holds true for buried seismic alarms, which can be triggered by thunderstorms, making inclement weather an excellent time to strike. Fence-mounted sensors are often triggered by heavy winds and snow. Such sensors are sometimes deactivated by lazy guards at the first sign of bad weather.

Some lighting systems are rigged to photoelectric cells, which turn the lights on whenever natural light dims. As a result, these lights switch off anytime a sufficiently bright flash of lightning tricks them. Night fighters can use artificial flashes and directed lasers to trick such lights into deactivating.

NVGs and other night observation devices (NODs) used by security personnel are practically useless on nights when lightning is flashing across the sky. Each flash has the ability to "white out" the equipment and render it temporarily useless.

Directional listening devices are at a loss during high winds and thunderstorms. When weather is amiable, a recorder with canned (recorded) static can confuse a directional mike.

Thermal imaging equipment can be circumvented by masking your heat signature with that of a large vehicle. When you are being tracked by a helicopter-mounted thermal imaging device, hide near large air-conditioning discharge vents or large generators whose output of heat will mask your own. The body heat from large animals, such as cattle, can also be used to mask your own body heat.

To recap, no matter how foreboding a perimeter may appear at first glance, there are flaws in every perimeter, fixed or flexible, that you can turn to your advantage.

- All infallible security alarms must be answered by fallible human beings.
- All electronic sensors rely on a vulnerable power supply.
- All alarm systems, at one time or another, give off false alarms.
- As fierce as guard dogs are, and as sophisticated and technologically advanced as computer monitors and sensors are, none can *think*.
- Any perimeter, no matter how well guarded or supported by what modern monitoring equipment, can be penetrated, given sufficient time.
- An installation can have row after row of perimeter security devices, but, ultimately, they depend on fallible human beings to monitor and respond to calls put out by those alarms. A sophisticated motion monitor might alert a security patrol to a perimeter breach, but, if on arrival, the security patrol finds a common house cat climbing on the sensitive fence, odds are that is as far as their investigation will go. It will probably never occur to them, tired and underpaid as they are, that the tabby might have been placed there by an infiltrator who is already inside the perimeter.

Finally, always approach a target in a zigzagging fashion. Practice disappearing into one shadow and reappearing from a

Illustration 25. The frontal zigzag approach.

I: Use shadows cast by objects, as well as the objects themselves, for cover.

II: When using objects for cover, mold your silhouette to fit their contours and lines.

III: Always keep your initial objective in sight. If that objective is a sentry in need of removal, do not stare at him straight on, which might alert his sixth sense. Instead, fix him in your peripheral vision, and beware of watching a single point or sentry to the exclusion of all else, which can make you vulnerable to discovery or attack.

different one without following a straight line. If your movement is too linear, an enemy who thinks he saw something out of the corner of his eye will tend to watch forward along the same line of movement. Breaking your movement pattern can cause him to lose track of you or doubt his initial observation.

STRUCTURES

Structures that must be penetrated or otherwise dealt with during an operation must be factored into any reconnaissance, because the types of structures as well as the lighting around them will greatly influence your ingress and egress, as well as your movements while inside the perimeter. In addition, there are certain methods of movement that must be followed while inside.

There are three areas of consideration when preparing to deal with structures: the exterior construction and defenses, the actual penetration, and how you move about while inside.

The Exteriors of Structures

You must take careful note of the colors of a building or walls you will be operating around; a light-colored backdrop can easily display your shadow or silhouette. The type of lighting around a structure can also betray you. Pass all windows, even those darkened, as if active (lighted). Never pass a window upright; always crouch under it. Remember that although the window may be darkened, someone entering the room could turn on a light, which would come through the window and catch you as you pass.

Shadows cast by structures are dependent on the turning on and off of both interior and exterior lighting (e.g., shining out through windows, from under doors). Beware of loose footing (e.g., gravel) outside of buildings.

The various conditions you must deal with when approaching a structure in the center of a guarded perimeter differ from those you are confronted with when approaching an isolated structure, such as a rural residence. The first step is to, whenever possible and without raising suspicion, disable all communications, such as telephone lines. (Unfortunately, the advent of cel-

Illustration 26. The rear zigzag approach.

I: To approach a sentry from the rear, move in a zigzag pattern until within three to five feet of him. Pause for a second and then rush across the remaining few feet to strike the target.

II: Always stay under the sentry's line of sight as you make your approach.

III: Approach the sentry with the sun or moon to your front to prevent your shadow from being spotted before you strike.

lular phones has made the total isolation of a house, building, or perimeter almost impossible.)

Any lighting that can be disabled without unduly arousing suspicion should be.

Penetrating a Structure

This can be done during the initial daylight reconnaissance in order to facilitate entry later that night. You will recall that this technique can be used on businesses and bases where visitors are permitted during the day, but where security is beefed up at night. Being able to walk in the front door during the day is always preferable to having to break glass or pry open a door in the dead of night, actions that could make noise or otherwise attract attention.

When looking to enter a structure, do not neglect basement or second-story windows, ventilation shafts, and access chutes used for coal, maintenance, and deliveries. Windows and openings above and below line of sight are given less attention than doors and windows at eye level.

When climbing up or down from a roof to enter second-story windows, or when climbing pipes or using a rope and grapple, always be careful not to leave footprints on the side of walls or kicked-in doors, which can be discovered by a passing sentry:

> As every detective knows, a criminal who might be meticulous in not leaving fingerprints is often very careless about his feet.
>
> —Frank Smyth
> *Cause of Death*

When entering or exiting a lighted room, use your jacket (or ninja cloak) to block the light that is flooding out. This should be done when opening any door. When entering through a window of a lighted room (something done only in an emergency), you reduce the size of your silhouette by pressing as close as possible to the frame so that your body blocks as little light as possible.

Methods of Movement within the Structure

You must beware of static from carpets and objects left on the floor (toys in a residence, golf balls in a businessman's office). Take care not to startle pets or let them startle you. (The presence of animals should have been discovered during any initial recon. This pertains not only to dogs but to other guard animals such as geese, guinea fowl, or parrots.)

Always be careful of vibrations you cause when walking across wooden floors or up wooden or metal stairs; someone at the top of the stairs or in an adjacent room might feel the vibrations. (This is especially true of the blind. In ancient Tibet, blind persons were

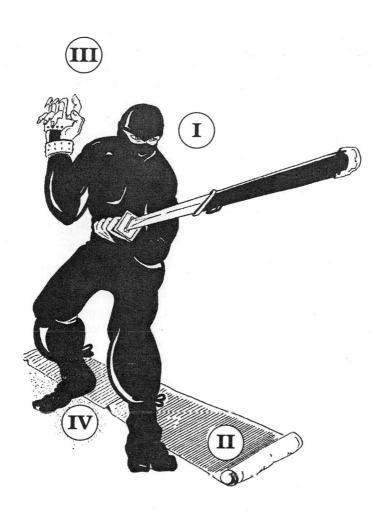

Illustration 27: The ninja method of moving through total darkness.

I: When moving through a pitch-dark room, ninja affixed their scabbards to the tips of their swords. An enemy touching this scabbard would seize it, placing him at the mercy of the true sword behind it.

II: The ninja walked on his sash to muffle his steps. This sash could also be rolled out and, when stepped on by an enemy, jerked out from under his feet. Today, this technique can be adapted for use with a rug, jacket, or blanket.

III: When unable to cross a room silently, ninja used nekode to move along rafters.

IV: Ninja always guarded their night passage by spreading rice or sand behind them (in a house) and by setting booby traps when moving through the bush.

often recruited into the sDop-sDop-trained bodyguards of kings and other notables. These sightless guards were perfectly at home in darkness and could detect night stalkers. Many of these blind sDop-sDop were master archers, capable of hitting targets by sound alone, and were unrivaled in close-quarter combat. Such guards were impossible to approach surreptitiously.)

In medieval Japan, samurai lords who were paranoid about ninja assassins had stairs and floorboards crafted specifically to make noise whenever someone stepped on them. As a counter, ninja developed a technique of unrolling a long sash on which they could walk silently. Another ploy used by samurai to protect themselves against creeping ninja was to spread rice in their hallways. Anyone stepping on this rice would inadvertently create a popping noise and betray their presence. Ninja countered this by moving along rafters, thus avoiding floors altogether. Ninja then added the rice technique to their arsenals, spreading rice behind them as they entered a structure to prevent enemies from sneaking up behind them. Modern criminal housebreakers often use this same ploy, spreading sand on linoleum, wood, or concrete floors. The crunching of this sand underfoot helps warn the burglar of someone's approach. In most instances, rice strewn on a floor would appear out of place, but sand would not because neglecting to wipe his feet could have dragged the sand in. A variation of this ploy is to place objects in front of the door leading into any room that you are working in. A rolling chair in an office or a child's toy in a residence can be strategically placed to

Illustration 28. Shadows on structures.

Figure one: For our structure facing north, shadows from the moon rising in the east obscures the west side, which remains in shadow until the moon climbs to midnight. As the evening progresses, shadows on the west side will get smaller by the hour.

Figure two: Nearing midnight, the moon directly overhead casts its light directly down on the structure. Except for a slight overhang of eaves, there will be no shadows around the structure (I). However, should the structure have prominent overhangs (a balcony or porch), the area under these overhangs will remain obscured in shadow (II).

Figure three: As morning approaches, the west side of the structure is now illuminated, while the east is shadowed. Beware the rising of the sun.

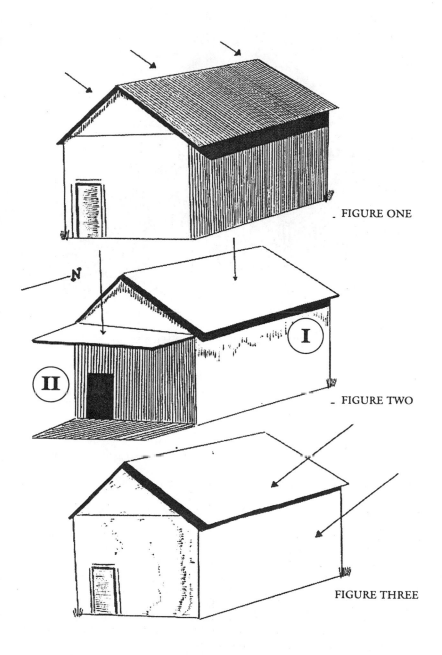

FIGURE ONE

N

I

II

FIGURE TWO

FIGURE THREE

Illustration 29. Shadows on structures.

The proximity of buildings to other buildings influences the interplay of light and shadows around both structures.

I: Provided that the moon is not directly overhead, one side of the structure will always be in shadow.

II: Light coming from inside a structure can destroy or distort shadows created by a structure.

III: Smaller and shorter structures remain in shadow when eclipsed by taller adjacent buildings.

IV: Smaller buildings produce their own shadows on sides facing away from adjacent buildings.

V: Shadows produced by this smaller building can also be influenced by light coming from its windows.

VI: Any area between two structures will form an alley of shadows, provided the moon (or sun) isn't directly overhead and light coming from inside structures does not destroy this corridor of darkness.

VII: Overhangs produce shadows of various sizes and angles while the moon or sun is rising or setting, and produce full shadows when the moon is directly overhead.

VIII: Trees and large shrubs cast shadows that can be taken advantage of. Trees have lots of leaves in summer and provide more cover in summer than in winter, when branches are bare. Nonetheless, bare branches, with their twisting and intertwining lines of silhouette, can also provide cover.

warn of another's approach. Burglars apply this principle by jamming toothpicks into a lock, causing anyone trying to unlock the door to have difficulty. The resulting noise alerts the burglar to the approach of danger.

Disable all lights when entering a structure to prevent them from being suddenly turned on. Familiarize yourself with all light switches, electrical cords, fuse boxes, and so on, because it may be necessary to plunge the room into darkness (or blind an enemy by suddenly turning on lights) to facilitate escape.

When entering a house or office, check for VCRs, clocks, and other timed devices that might suddenly go off during your foray, startling you. This also applies to phones.

Before entering a room, flash it using your afterimaging technique to ensure no one is lying in wait in the room.

Working in Total Darkness

In most situations there will be some degree of illumination. If that light doesn't exist in the environment itself, it can usually be brought into the environment in the form of flashlights or NODs. Medieval ninja were masters of tactics and techniques that allowed them to move more safely through pitch darkness.

When moving through a pitch-black room with an enemy, drop into a crouching or crawling position to prevent him from hitting you if he sprays the room with gunfire. When an enemy does fire a gun, pinpoint his position by his muzzle flashes. **Warning**: Do not look directly toward where you think a gun-toting enemy is hiding in the dark, since his muzzle flashes may temporarily destroy your night vision. When firing your own gun in the dark, the rule is: Fire and move, fire and move.

If you are carrying rice or sand in your pockets, scatter it toward your enemy in a wide swath at eye level. This serves two purposes. First, it will startle and further blind him if it hits him, thus revealing his position. Second, rice or sand falling on the floor near him will give away his position when he steps.

When moving through a dark room, move along walls whenever possible.

When working alone, never exit a structure the same way

you went in. You are always in more danger during egress than during ingress because of the fact that you may have been heard moving about inside.

Depending on the time of day or night, the sun or moon casts shadows differently. Factor in the hourly movement of shadows into any reconnaissance.

CHAPTER ELEVEN

TOOLS OF THE TRADE

You're only as good as your training and your tools. Your training consists of gallons of sweat spilled during the course of mastering the art of stealth. Whereas it has been said that the more you sweat in peace, the less you bleed in war—an undeniable truth—the most important part of any job is learning to recognize the right equipment needed to do that job.

Choosing the correct tools and weapons to accomplish your objective and overcome any obstacles depends on your initial reconnaissance of that objective and your ability to recognize those obstacles. The types of terrain you will be operating in, the types of obstacles you anticipate having to overcome, and your chosen method of ingress and egress all affect your choice of equipment.

All gear falls into one of five categories: clothing, bionics, traverse equipment, communications equipment, and weapons.

CLOTHES

Clothes make the man, or, in this case, clothes keep the man alive! Clothes must be chosen with two considerations in mind: fit and function. Clothes worn by the night fighter must fit the circumstances. For example, it would be foolish to wear "chocolate chip" desert camouflage in a snowy environment. Likewise, snow-white alpine camouflage would be out of place in a jungle, and green bush camouflage might prove a liability in an urban AO where sedate street clothes would prove more appropriate.

The clothes you wear must also be functional in order to deal with the elements, help shield you from discovery, and help you accomplish your objective.

> Ninja have always known the importance of choosing function over fashion. Fashions come and go, survival is always in style!
>
> —Dirk Skinner
> *Street Ninja*

For urban operations, a reversible hooded jacket might fit the bill. Coveralls can keep street clothes hidden and clean of crime scene fibers. Pants can be used as a flotation device in an emergency water crossing. Footwear must be chosen with care, since tracks left behind (in mud, on a carpet, on that wall that you climbed, or on a door you kicked in) can tell a lot about you.

> Sneakers, baseball boots and other canvas-and-rubber shoes tend to be among the favorite working footwear of such criminals as cat burglars because of their lightness and non-slip qualities. Unfortunately for their users, such shoes have drawbacks, too: canvas is easily snagged, leaves fibers at the scene of the crime, and it also marks easily, giving investigators the opportunity to compare stains on the shoe with substances on the entered premises.
>
> —Frank Smyth
> *Cause of Death*

From footprints alone, an expert can tell your gender, height, weight (from the depth of the print), height (from the distance between steps), and even certain aspects of your health (e.g., limp).

For night fighters, especially those who expect to come up against security forces, attention to clothing is especially important in a forensic sense; police and other security forces collect fibers left at the scene of a break-in and, once a suspect is apprehended, from fibers picked up by his clothing while at the scene. As early as 1884, detectives understood how important the type of clothing worn by criminals was.

> When entering a room, the thief always dresses in soft woolen clothing, and wears woolen stockings upon his feet. The noise made by the rustling of a muslin or linen shirt, when everything is hushed and still, is often sufficiently loud or harsh to awaken people, particularly ladies, from a sound and comfortable slumber.
>
> —Allen Pinkerton
> *Thirty Years a Detective*

Choose your clothing with your objective in mind. If climbing is involved you need footwear that will aid in your ascent and descent and not leave marks on walls. In fording water, dress in warm, dry clothing after your crossing. Either cross naked or secure a change of clothes in a waterproof bag.

BIONICS

Bionics consist of devices that augment your senses, such as NODs and directional microphones. NODs come in two basic varieties: infrared, which send out a beam of infrared light, and light enhancement, such as "starlight" scopes, which function in the same way a cat's eyes do by magnifying present light.

Infrared scopes and goggles can be used in total darkness, since they generate their own (infrared) light source. Their major drawback is that their use restricts both depth perception and peripheral vision.

Starlight scopes are convenient because they can be mounted on a rifle or machine gun. However, these scopes cannot operate in total darkness, because they rely on magnifying existing light rather than producing their own.

Noise amplifiers and directional mics help to pick up sounds that the human ear either cannot hear or hears poorly. Often such devices are hooked to alarms. Their main shortcoming is that, even though the device can pick up sound, a human being must still decipher friend from foe.

Although bionics have their place in your arsenal, there is always the danger of relying on such devices to the point that you allow your natural senses to atrophy. You must familiarize yourself with these devices, if only to avoid becoming victims of them.

In early 1992, the Oregon Court of Appeals ruled that police could not use night-vision scopes without a search warrant, since such scopes enable police to see what they can't see with the naked eye.

TRAVERSE EQUIPMENT

Traverse equipment consists of any vehicles or specialized gear needed to get you into and out of your AO. This includes any special equipment needed while you are in your AO.

To cross bodies of water, you may need rafts, scuba gear, or even a minisub. Knights of darkness use various kinds of vehicles, both legal and commandeered. These include motor vehicles, boats, and aircraft. Specialized terrain calls for special traverse equipment, such as snowmobiles, ATVs, and even pack animals. It could consist of something as mundane as getting bus or train schedules for an urban AO. If the objective is situated near or in mountains or tall structures, you might need to acquire parachutes. In a city or in the mountains, you may need climbing and rappelling equipment.

COMMUNICATIONS EQUIPMENT

This consists of walkie-talkies and radios, all fitted with earphones to avoid static, which could give you away. Other forms of communication include mirrors, sign language, tap codes, animal sounds, and even smoke signals.

WEAPONS

Weapons are an important component of any night operation.

Medieval ninja became experts at incorporating weapons into their clothing and in using the clothing itself as a weapon, since they were required, as are most night fighters, to travel light. The ninja sash could be used as a climbing aid, garrote, or binding to tie up an enemy. Ninja face scarves not only helped disguise faces but also to muffle breathing. Sewn into the elbows and knees of their clothing were spikes that aided in climbing and close-quarter combat. We've already mentioned how cloaks worn by such night fighters as ninja and wolfshirts could be used to disguise their silhouettes. Cloaks could also be thrown over an enemy's head or used to pull an enemy's feet out from under him.

Your clothing, including your street clothing, should be chosen for utility rather than fashion. In his *Death on Your Doorstep: 101 Weapons in the Home* (Alpha Publications of Ohio, 1993), Ralf Dean Omar lists dozens of ways that everyday objects of clothing can be used for defense and offense.

If your objective is to gather intelligence, you will carry fewer weapons, since the nature of the operation requires getting in and getting back out fast while avoiding contact with enemy troops. However, if your objective is to penetrate an enemy perimeter for purposes of sabotage or assassination, you will carry a considerable arsenal of specialized weapons and explosives.

One major consideration for choosing your weapons will be the type of weapons your enemy carries. Carrying similar weapons allows you to better fit in (e.g., silhouette) and to borrow ammunition when needed. In a night fight, using the same type of weapons as your enemy makes it harder for an enemy to identify you by a distinctive muzzle flash or the sound of your weapon.

Because no gun can be made completely silent, firearms have limited use in dispatching sentries. In addition, the sound

WEAPON	DISTANCE	NOISE	ADVANTAGE	DISADVANTAGE
Silenced firearm	Contact to many yards	1-5	Efficiency, distance	Muzzle flash, spent casings, ballistics trace
Stealth guns (spring-fired)	Contact to several yards	1-5	Ease of use	Low impact
Stealth guns (CO_2-fired)	Contact to several yards	1-5	Impact, darts can be poisoned	Low impact
Bludgeon	Contact	1-5	Environmental	Requires contact with enemy
Bow & arrow	Yards	1	Cuts through bulletproof vest	Time to master, most archery not combat oriented, bulky
Crossbow	Yards	1-2	Knockdown power, ease of use	Bulky
Garotte (wire, stick, arm)	Contact	1-3	Multiple methods, environmental	Requires contact with enemy, time to master
Blade (held)	Contact	1-3	Doubles as a club	Requires contact with enemy
Blade (thrown)	Few feet	1-5	Bridges distance	Low impact, loss of weapon
Blade (shooting knife)	Few feet	1-5	Hilt doubles as a bludgeon	Lack of accuracy, single shot
Shuriken (ninja star)	Contact to yards	1-3	Poisoned, close combat	Harassment value only, unless poisoned
Taser (shooting stun gun)	Contact to two yards	5-10	Ease of use	Noise, flash of electricity
Stun gun (hand-held)	Contact	1-5	Ease of use, efficiency	Possible buzzing noise, flash of electricity
Spear (atlatl)	Yards	1-3	Impact	Time to master, bulky
Spear (thrown)	Yards	1-3	Impact	Time to master, bulky
Spear guns (aquatic)	Yards	1-5	Ease of use, penetration	Accuracy on land
Body blows	Contact	1-5	Efficiency	Time to master, require contact with enemy
Blowgun	Yards	1-5	Poison darts	Time to master, low impact

Illustration 30: Stealth weapons assessment.
(Note: This refers to the potential level of noise,
with 1 being the lowest and 10 being the highest.)

of an automatic firearm's slide being ratcheted, the smell of gun oil and gunpowder, and spent casings left scattered about can all expose you. In place of traditional firearms, several alternative "stealth" weapons are available to you, all with varying levels of effectiveness.

For example, several types of stealth guns are available. Some of these are spring operated, and others fire a projectile by CO_2 (gas compression). Others use springs or rubber cords to propel fletchettes. But, as with firearms, no stealth gun is completely silent. Most make at least a popping or whooshing sound. Other stealth weapons include guns that fire stun bullets (rubber bullets or tiny beanbags) or which use electricity to disable an enemy. Still, beanbags and rubber bullets are seldom silent, either. Stun guns such as the Taser (which fires electric prongs that remain attached to the shooting device) also fall short in the silence department. Further, small, hand-held stun guns, when pressed into the neck muscles, immediately incapacitate a sentry. However, the drawbacks are the flash of electricity that appears when the gun's contacts fire and the audible electric "buzz" heard when the gun is activated.

Conventional, spring-loaded aquatic spear guns can be useful under certain circumstances, but they do make noise. Bows have long been favored and specialized lightweight collapsible bows originally designed for special forces are now available. Crossbows, too, are excellent weapons for stealth operations. (Any noise from the "twang" of a crossbow's firing is offset by a crossbow's knockdown power.) Spear and arrow weapons have the added advantage of being able to penetrate bullet-resistant vests.

Knives, which at close quarters can be very effective, can also be used if the handler is properly trained and uses the right knife. For instance, a knife should never be thrown at an enemy, because even hitting your foe with a thrown blade is no guarantee that he will be silenced (even when dying, a sentry can sound the alarm). Knives specially designed for throwing do exist, however. Perhaps the best known of these is the *shuriken*, which is also known as the ninja star. Traditional ninja stars were used to dissuade pursuit. Unless poisoned, a small shuriken has harassment value only. However, in a pinch, a shuriken held like a razor blade can be used to slash a throat. One viable modern variation of the shuriken consists of two four- to eight-inch stainless-steel blades that can be carried closed and then opened and locked in place to form a four-pointed "X" throwing tool. It is almost impossible for this X-knife not to stick into a target.

Soviet Spetsnaz developed a "shooting knife" whose blade could be propelled a distance of several feet. Originally these blades were spring-loaded, but later imitators use compressed CO_2. For additional information on edged weapons, see my *Assassin! The Deadly Art of the Cult of the Assassins.*

You can also use a length of cord, wire, or stick to garrote an enemy. (See *The Ancient Art of Strangulation.*) During the Korean War, South Korean night fighters wrapped protective wire mesh around their throats (protection against strangulation) and their lower backs (to guard against rear knife attacks). Never assume that you are the only predator in the night.

Bashing a sentry over the head with a heavy bludgeon (rock, heavy stick) is also effective, as are certain hand blows and arm strangles. In the final analysis, the body is always your first weapon and your ultimate line of defense.

Fancy thinking the Beast was something you could hunt and kill.

—William Golding
Lord of the Flies

Long before the advent of the incandescent bulb and before neon cast its otherworldly glow over the morals of man, a desperate Egyptian pharaoh decided to try to banish darkness and shadows from his realm with the light of bonfires. Amahotep IV, also known as Akhenaton, decreed that a city should be built, a city whose great towers and temples would be crafted to such perfection that no wedge of shadow, no outline of dark silhouette would be found anywhere within its confines. The city was to be a monument to Ra, the god of the sun.

Akhenaton never lived to see his project completed; he was assassinated shortly after construction began on his shadowless city by priests of the old gods who were at odds with the young pharaoh's radical monotheism. If we are to judge Akhenaton, we must conclude that his fault was not that he dreamed of enlightening the world, nor that he wished for his subjects to live in the light rather than dwell in their fear of skulking shadows, disembodied silhouettes, and those black things of chaos and corruption that have vied for the sanity and the soul of man since time immemorial. Indeed, Akhenaton's shortsightedness lay not in his golden vision nor in his failure to recognize that too much light blinds as surely as too little. No, Akhenaton's failure is to be found in his inability to recognize that, no matter what the intensity of any artificial light man's insight or fear creates to hold the night at bay, the shadows that man has most to fear wait ever within his own heart of darkness.

Bibliography and Suggested Reading

Cruickshank, Charles. *Deception in World War II*. London: Oxford University Press, 1979.

Dachman, Ken and John Lyons. *You Can Relieve Pain: How Guided Imagery Can Help You Reduce Pain or Eliminate It Altogether*. New York: Harper & Row, Publishers, 1990.

Daraul, Arkon. *A History of Secret Societies*. New York: Citadel Press, 1962.

Kipper, David. "Stress for Success." *Prime* (Summer 1996): pp. 34–36.

Lovejoy, David. "Night Ordeal." *Gallery* (June 1993): pp. 106–108.

Lung, Dr, Haha. *The Ancient Art of Strangulation*. Boulder: Paladin Press, 1995.

——— *The Ninja Craft*. Ohio: Alpha Publications of Ohio, 1996.

——— *Assassin! The Deadly Art of the Cult of the Assassins*. Boulder: Paladin Press, 1997.

Omar, Ralf Dean. "Ninja Death Touch: The Fact and the Fiction." *Black Belt* (September 1989).

——— *Death On Your Doorstep: 101 Weapons in the Home.*

Ohio: Alpha Publications of Ohio, 1993.

Allen Pinkerton. *Thirty Years a Detective*. New York: G.W. Carlton & Company, 1884.

Frank Smyth. *Cause of Death: The Story of Forensic Science*. New York: Van Nostrand Reinhold Company, 1980.

Sun Tzu. *Ping Fa* (*The Art of War*) (multiple translations).

About the Author

Dr. Haha Lung is the author of more than a dozen books dealing with the darker side of the martial arts world, including *The Ancient Art of Strangulation* and *Assassin! The Deadly Art of the Cult of the Assassins.*